Animal Trivia

A Massive Collection of Animal Fun Facts, Anecdotes, and Pub Quiz Trivia

Animal Trivia

A Massive Collection of Animal Fun Facts, Anecdotes, and Pub Quiz Trivia

Copyright 2024
Adicus Abbott

Have fun with this collection of animal trivia. The information in this book of trivia is for entertainment purposes only.Enjoy!

ISBN: 9798340046444
Independently Published

Table of Contents

Welcome to Animal Trivia   7

Mammals   15

Insects   39

Arachnids   59

Birds   81

Reptiles   101

Amphibians   123

Fish   147

Crustaceans   171

Mollusks   193

Wee Beasties   215

Welcome to Animal Trivia

Before we get started, I want to take you back to your 8th grade biology class where you learned about phylums and kingdoms. I know. If you're like me, you've forgotten more than you ever learned about this subject.

But this is important stuff. In a book dedicated to animal trivia, it's helpful, in my opinion, to first understand where animals fit in the world of living organisms. And if you're ready to get geeky up in here, it all begins with a thing called, Linnaean Taxonomy.

Linnaean Taxonomy

Linnaean taxonomy is a system that scientists use to classify and organize all living things on Earth. It was created by Carl Linnaeus, a Swedish botanist, in the 18th century. This system helps scientists understand how different organisms are related to one another and provides a way to name them in a clear and consistent manner.

In science, as in any pub quiz trivia type competition, accuracy is important.

Imagine a pyramid. At the top is a massive stone called Domain. Smaller stones spread out under this capstone, all the way to the base, where you find species. Each level is more specific than the last. Let's break down the different levels of Linnaean taxonomy and see how they work.

Domain

The highest level of classification is the domain. There are three domains of life: Bacteria, Archaea, and Eukarya. Bacteria and Archaea are made up of single-celled organisms without a nucleus, while Eukarya includes all organisms whose cells have a nucleus, such as plants, animals, fungi, and protists.

Kingdom

Within each domain, life is further divided into kingdoms. For example, the domain Eukarya has four major kingdoms: Animalia (animals), Plantae (plants), Fungi (mushrooms, yeast, etc.), and Protista (mostly single-celled organisms like algae). The kingdom is one of the largest groups and contains millions of different species.

Phylum

Each kingdom is divided into smaller groups called phyla (the plural of phylum). In the animal kingdom, one example is the phylum Chordata, which includes animals that have a notochord at some point in their life. This group includes mammals, birds, fish, and reptiles. Plants and fungi also have their own phyla, based on specific characteristics unique to them.

Class

Next, each phylum is divided into classes. For example, in the phylum Chordata, there is the class Mammalia, which includes all mammals. Mammals are warm-blooded animals with hair or fur and have the ability to nurse their young with milk. Classes further narrow down the types of organisms into more specific groups based on shared traits.

Order

Each class is then broken down into orders. For instance, within Mammalia, there is the order Carnivora, which includes meat-eating mammals like lions, bears, and wolves. Orders help scientists focus on organisms that are even more closely related to one another.

Family

Within each order, organisms are grouped into families. Continuing with our example, within the order Carnivora, there is the family Felidae, which includes all cats, both big and small, like lions, tigers, and domestic cats.

Genus

Next comes the genus. The genus includes species that are very similar to each other. In the family Felidae, for example, the genus Panthera includes big cats like lions, tigers, leopards, and jaguars. Genus names are always written in italics and with a capital letter.

Species

Finally, the most specific level of classification is the species. A species includes organisms that can breed and produce fertile offspring. For example, the species Panthera leo refers to lions, while Panthera tigris refers to tigers. The species name is written in italics and is always lowercase.
Importance of Linnaean Taxonomy

Linnaean taxonomy is important because it allows scientists to identify and name organisms in a universal way. It also helps us understand how different species are related to each other, from the largest groups down to individual species. This system is used all over the world, ensuring that scientists can communicate clearly about life on Earth, no matter where they are from.

But the focus of this book is animal trivia.

The Animal Kingdom

Okay. Are you ready for this? We're getting ready to step into some deep weeds here.

Within the Linnaean taxonomy, animals are classified based on shared characteristics into different groups, starting with the kingdom Animalia. As you move down the taxonomic levels, animals are further divided into various phyla, classes, orders, families, genera, and species. Below are some of the major types of animals, focusing on the phylum and class levels within the animal kingdom.

Major Phyla in the Animal Kingdom

Phylum Chordata: This phylum includes animals with a notochord, which is a flexible, rod-like structure found at some point during development. The most well-known groups of animals, such as mammals, birds, reptiles, amphibians, and fish, belong to this phylum.

- Class Mammalia (Mammals): Warm-blooded animals with hair or fur and the ability to nurse their young with milk. Examples include humans, dogs, cats, and whales.
- Class Aves (Birds): Warm-blooded, feathered animals that lay eggs. Examples include eagles, sparrows, penguins, and ostriches.
- Class Reptilia (Reptiles): Cold-blooded animals with scaly skin. Many lay eggs, and examples include snakes, lizards, turtles, and crocodiles.
- Class Amphibia (Amphibians): Cold-blooded animals that live part of their life in water and part on land. They typically have moist skin, and examples include frogs, salamanders, and toads.
- Class Osteichthyes (Bony Fish): This class includes fish with bony skeletons, like salmon, trout, and goldfish.
- Class Chondrichthyes (Cartilaginous Fish): These are fish with skeletons made of cartilage instead of bone. Sharks, rays, and skates are examples.

Phylum Arthropoda: The largest phylum in the animal kingdom, arthropods have exoskeletons, segmented bodies, and jointed legs. This phylum includes insects, spiders, and crustaceans.

- Class Insecta (Insects): Insects have three body segments (head, thorax, and abdomen), six legs, and often wings. Examples include ants, butterflies, and beetles.
- Class Arachnida (Spiders and Scorpions): Arachnids have eight legs and two body segments. This class includes spiders, scorpions, ticks, and mites.
- Class Crustacea (Crustaceans): These are mostly aquatic animals with hard exoskeletons. Examples include crabs, lobsters, shrimp, and barnacles.

Phylum Mollusca: Mollusks are soft-bodied animals, often with a hard shell. This phylum includes animals like snails, clams, and squids.

- Class Gastropoda (Snails and Slugs): These mollusks usually have a single, spiraled shell (or no shell in the case of slugs). Examples include garden snails and sea slugs.
- Class Cephalopoda (Squids and Octopuses): These animals are highly intelligent and have tentacles. Examples include octopuses, squids, and cuttlefish.
- Class Bivalvia (Clams, Oysters, Mussels): These mollusks have two-part shells and include clams, oysters, and mussels.

Phylum Echinodermata: Echinoderms are marine animals with radial symmetry and a water vascular system. They include sea stars, sea urchins, and sand dollars.

- Class Asteroidea (Sea Stars): These animals have a star-shaped body and are known for their ability to regenerate lost arms.
- Class Echinoidea (Sea Urchins and Sand Dollars): These animals have a round, spiny body or a flattened body, like sand dollars.

Phylum Annelida: These are segmented worms, including earthworms and leeches.

- Class Clitellata: This includes earthworms and leeches, animals that have a segmented body with no hard exoskeleton.

Phylum Cnidaria: These are aquatic animals that have specialized stinging cells called cnidocytes. This phylum includes jellyfish, corals, and sea anemones.

- Class Scyphozoa (Jellyfish): Jellyfish are free-swimming animals with tentacles that have stinging cells.
- Class Anthozoa (Corals and Sea Anemones): These animals are usually stationary and form coral reefs or live as solitary polyps.

I know. Some of the critters listed above are beyond odd. So, for the purposes of this book, let's narrow the scope of our discussion to:

- Mammals
- Insects
- Arachnids
- Birds

- Reptiles
- Amphibians
- Fish
- Crustaceans
- Mollusks
- Wee Beaties

Sweet. I can wrap my arms around that. Let's get started.

Mammals

Mammals are a type of animal that share a few key features. First, all mammals are warm-blooded, which means they can keep their body temperature steady, no matter how hot or cold it is outside. Second, mammals have hair or fur at some point in their lives, which helps keep them warm. Another special trait is that female mammals have mammary glands, which produce milk to feed their babies.

Mammals come in all shapes and sizes, from tiny mice to massive whales. Some live on land, like dogs, cats, and humans, while others, like dolphins and seals, live in the water. Mammals are also unique because most give birth to live young instead of laying eggs, although there are a few exceptions, like the platypus and echidna, which do lay eggs.

Most mammals also have lungs and need to breathe air, and they have a backbone to support their bodies. Many mammals are known for their high intelligence, and some, like humans, dolphins, and chimpanzees, are capable of complex thinking and social behavior.

In short, mammals are warm-blooded animals with fur or hair, most give birth to live young, and they nurse their babies with milk.

Now, before we get into the questions and answers regarding mammals, here are some fun facts about one of our favorite mammals, the house cat.

- Flexible Spines: Cats' spines are incredibly flexible, allowing them to twist and turn mid-jump, giving them agility and balance.
- Whisker Sensitivity: Cats' whiskers are highly sensitive and help them navigate tight spaces by detecting changes in the air.
- No Sweet Tooth: Cats lack taste receptors for sweetness, so they don't experience sugary flavors.
- Purring Benefits: Cats purr not only when they're happy but also to self-heal. Purring frequencies can promote tissue regeneration and bone healing.
- Unique Walk: Cats walk by moving both left legs, then both right legs, a trait shared only with camels and giraffes.
- Night Vision: Cats can see in one-sixth of the light humans need, thanks to extra light-reflecting cells in their eyes.
- Fast Reflexes: Cats' reflexes are faster than most animals, with the ability to react in less than a second.
- Sleeping Experts: Cats sleep 12-16 hours a day, conserving energy for bursts of activity.
- Grooming Obsession: Cats spend up to 30% of their time grooming to regulate body temperature and remove scent traces.

Not to leave out the dog lovers among us, here are some interesting dog fun facts.

- Sense of Smell: A dog's sense of smell is around 40 times better than a human's, with some breeds able to detect scents from miles away.
- Unique Nose Prints: A dog's nose print is as unique as a human fingerprint, making it possible to identify them by their nose patterns.
- Wide Range of Hearing: Dogs can hear frequencies between 40 Hz and 60,000 Hz, which is much higher than humans. This helps them pick up distant sounds.
- Dreaming Pups: Like humans, dogs experience REM (rapid eye movement) sleep and dream, often twitching or moving in their sleep.
- Sweat Through Paws: Dogs don't sweat through their bodies like humans. Instead, they release heat through the pads of their paws.
- Dog Communication: Dogs communicate through a combination of body language, vocalizations (barks, growls), and scent marking.
- Tail Talk: A dog's tail movement can indicate mood. A wag to the right suggests happiness, while a wag to the left may indicate anxiety.
- Super Athletes: Some dogs, like Greyhounds, can run up to 45 mph, making them among the fastest animals on land.
- Social Pack Animals: Dogs are pack animals and thrive on companionship, making them deeply loyal to their owners.

Okay, let's get this party started with a deep dive into a few dozen…or more…pub quiz style question and answer trivia fun facts and anecdotes.

Q: What is the largest mammal on Earth? A: The blue whale.

Q: Which mammal is known for laying eggs? A: The platypus and echidna.

Q: What is the only mammal that can truly fly? A: The bat.

Q: Which mammal has the longest pregnancy? A: The African elephant, with a pregnancy lasting up to 22 months.

Q: Which mammal has the largest brain in relation to its body size? A: The human.

Q: Which land mammal has the thickest fur? A: The sea otter.

Q: What mammal has the longest lifespan? A: The bowhead whale, which can live over 200 years.

Q: Which mammal can hold its breath underwater for the longest time? A: The sperm whale, for up to 90 minutes.

Q: Which mammal is known to sleep the most hours per day? A: The koala, which sleeps up to 22 hours a day.

Q: Which mammal has fingerprints almost identical to humans? A: The koala.

Q: What is the fastest land mammal? A: The cheetah, reaching speeds up to 70 mph.

Q: What is the smallest mammal by weight? A: The Etruscan shrew.

Q: Which mammal is known for having pink milk? A: The hippopotamus.

Q: Which mammal can turn its head almost 180 degrees? A: The owl monkey.

Q: Which mammal is immune to snake venom? A: The mongoose.

Q: Which mammal spends almost 90% of its life in trees? A: The sloth.

Q: What is the only mammal to live on every continent? A: Humans.

Q: Which mammal is capable of echolocation, besides bats? A: Dolphins.

Whales and dolphins are seriously smart, and their brains are as impressive as you'd expect for creatures that live such complex social lives. Both belong to a group called cetaceans, and they have some of the largest brains in the animal kingdom. Dolphins, especially, are known for their high intelligence. They have large brains relative to their body size, with highly developed areas for problem-solving, communication, and social interaction.

Dolphins are often seen working together in groups, or pods, and they use a variety of clicks, whistles, and body movements to communicate. They can recognize themselves in mirrors (a sign of self-awareness), use tools (like covering their noses with sponges to protect themselves while foraging), and even understand complex commands when trained by humans.

Whales, particularly species like orcas (killer whales), are also incredibly intelligent. They have their own unique vocal dialects, passed down through generations, and exhibit behaviors that suggest culture—like teaching younger whales how to hunt or use specific strategies to catch prey.

Both whales and dolphins also show signs of empathy and grief, which are hallmarks of advanced cognition. So, while they may not be cracking open math textbooks, they're clearly some of the brainiest creatures in the ocean!

Q: What mammal has the largest eyes relative to its body size?
A: The tarsier.

Q: Which mammal has the longest tail relative to its body size?
A: The giraffe.

Q: Which mammal holds the record for the most teeth? A: The giant armadillo, with up to 100 teeth.

Q: Which mammal has the largest mouth in the animal kingdom? A: The blue whale.

Q: Which mammal is known for using tools? A: The chimpanzee.

Q: Which mammal is the heaviest land animal? A: The African elephant.

Q: Which mammal has no vocal cords? A: The giraffe.

Q: Which mammal has the shortest pregnancy? A: The Virginian opossum, with a gestation period of just 12 days.

Q: Which mammal has the most bones in its spine? A: The manatee, with up to 60 vertebrae.

Q: What is the only mammal with a shell? A: The armadillo.

Using armadillo meat in chili might sound unusual, but in some regions, especially in parts of the southern United States and Latin America, it's been done! Armadillo meat is lean and has a flavor similar to pork, making it a unique addition to a pot of chili. It adds a bit of gamey richness that some chili lovers swear by. Like other wild game meats, armadillo should be cooked thoroughly to avoid any health risks, but when done right, it blends well with chili's spices, beans, and tomatoes, creating a hearty, flavorful dish. Definitely a bold choice for adventurous eaters!

Q: Which mammal never stops growing throughout its life? A: The elephant.

Q: What is the only mammal that lives entirely underground? A: The naked mole-rat.

Q: Which mammal has the strongest bite force? A: The hippopotamus.

Q: Which mammal has opposable thumbs, other than humans? A: The lemur.

Q: What is the only mammal with wings? A: The bat.

Q: Which mammal has the densest bones? A: The whale.

Q: Which mammal has the most complex social structure? A: The elephant.

Q: What is the largest rodent in the world? A: The capybara.

Q: Which mammal is known to "laugh" when tickled? A: The rat.

Q: Which mammal is the most widespread terrestrial carnivore? A: The red fox.

Q: Which mammal has the shortest lifespan? A: The house mouse, living about 1-2 years.

Q: Which mammal is known for playing dead when threatened? A: The opossum.

Q: What is the tallest land mammal? A: The giraffe.

Giraffes are tall for a pretty simple reason: they evolved that way to reach food that other animals can't. Their long necks give them access to the leaves and branches of tall trees, especially acacias, which are a favorite. While other herbivores are stuck munching on grass or low shrubs, giraffes can go straight to the top and get all the good stuff without competition.

Interestingly, their height is also helpful in spotting predators like lions from a distance. Being tall gives them a great view of the savanna, which helps them stay alert. There's also a bit of a romantic twist—male giraffes use their long necks to fight for mates in a behavior called "necking." They swing their heads like clubs to knock each other around, and the taller, stronger one usually wins.

So, being tall isn't just about food—it helps giraffes survive and thrive in more ways than one!

Q: Which mammal has the most complex brain aside from humans? A: The dolphin.

Q: What mammal is sometimes called the "sea cow"? A: The manatee.

Q: Which mammal is known for migrating the longest distance? A: The gray whale.

Q: What mammal has the largest land-based migration? A: The caribou.

Q: Which mammal has the largest brain of any animal? A: The sperm whale.

Q: What is the only mammal that can hibernate for months? A: The bear.

Q: Which mammal can jump the highest relative to its size? A: The flea.

Q: Which mammal is known for building dams? A: The beaver.

Q: What is the only marsupial native to North America? A: The opossum.

Q: Which mammal can rotate its rear feet 180 degrees to climb down trees headfirst? A: The squirrel.

Q: What mammal is known for its incredible sense of smell, often used in truffle hunting? A: The pig.

Q: Which mammal has the longest tongue relative to its body? A: The anteater.

Q: What mammal is the largest carnivore on land? A: The polar bear.

Q: Which mammal spends almost its entire life in the water? A: The whale.

Q: What mammal has retractable claws? A: The cat.

Q: Which mammal is known for "surfing" on waves while hunting? A: The orca (killer whale).

Q: Which mammal has the best night vision? A: The owl monkey.

Q: What mammal has the strongest sense of hearing? A: The bat.

Q: Which mammal can recognize itself in a mirror? A: The elephant.

Q: What is the largest species of bear? A: The Kodiak bear.

Q: Which mammal has the most advanced form of communication, besides humans? A: The dolphin.

Q: What mammal can swim the fastest in water? A: The common dolphin.

Q: Which mammal has the most fur per square inch? A: The sea otter.

Q: Which mammal is known to cry real tears? A: The elephant.

Q: What is the slowest moving mammal? A: The three-toed sloth.

Q: Which mammal has webbed feet for swimming? A: The beaver.

Q: What mammal is famous for its "laughing" sound? A: The hyena.

Q: Which mammal is known to mate for life? A: The beaver.

Beavers build dams for a pretty smart reason—they're creating their own safe, watery home! Beavers are excellent swimmers, but they're pretty clumsy on land. By building dams, they flood areas to create ponds, which provide them with protection from predators like wolves, bears, and coyotes. In the middle of these ponds, they build their cozy lodges, where they can live and raise their young in safety.

Another reason beavers build dams is to store food. In the fall, they stockpile branches and sticks underwater near their lodge, so they have plenty to eat during the winter when the pond is frozen.

The flowing water of rivers and streams can be too fast for their lodges, so they build dams to slow the water down and create a calmer, more stable environment. It's nature's version of home improvement—beavers are making sure they have a safe, stable spot to live and thrive!

Q: Which mammal uses sonar to navigate and hunt in the dark? A: The bat.

Q: What mammal has the largest number of species? A: The rodent.

Q: Which mammal is known for building underground tunnels? A: The mole.

Q: Which mammal has tusks and lives in cold climates? A: The walrus.

Q: Which mammal sleeps standing up? A: The horse.

Q: What mammal's heart beats the slowest? A: The blue whale, with as few as 2 beats per minute.

Q: Which mammal has the best memory? A: The elephant.

Q: What mammal is the only one to have been domesticated from the camelid family? A: The alpaca.

Q: Which mammal can regrow its teeth multiple times throughout its life? A: The dolphin.

Q: Which mammal is famous for living in the trees of Madagascar? A: The lemur.

Q: What mammal can sleep with half its brain awake? A: The dolphin.

Ready to up the action? Here are a few more advanced questions.

Advanced Fun Facts

Q: What is the only mammal with knees that bend backward? A: The elephant.

Q: Which mammal's metabolism allows it to eat up to 20 times its body weight each day? A: The shrew.

Q: What mammal has the most extended migration of any mammal based on straight-line distance? A: The humpback whale.

Q: Which mammal has a specialized organ called the "melon" used for echolocation? A: The dolphin.

Q: Which mammal has the lowest known metabolic rate? A: The three-toed sloth.

Q: What species of mammal is known for having the longest gestation period of any carnivore? A: The orca (killer whale), with a gestation period of up to 17 months.

Q: Which mammal has the largest number of vertebrae in its neck? A: The manatee, with six cervical vertebrae, whereas most mammals have seven.

Q: Which mammal's skin is so thick it can withstand bites from lions and crocodiles? A: The hippopotamus.

Believe it or not, hippos are considered the most dangerous animals in Africa! While they might look like slow, lumbering giants, they're surprisingly fast and aggressive when they feel threatened. Weighing up to 3,000 pounds and with huge, powerful jaws, hippos are responsible for more human deaths in Africa than lions, crocodiles, or any other wildlife.

Hippos are incredibly territorial, especially when it comes to their water sources. If you get too close to their rivers or lakes, they'll charge with surprising speed—on land, they can run up to 20 mph! In the water, they're even more dangerous, as they can easily tip boats or attack anything they see as a threat.

Despite being herbivores, they have massive tusks that can deliver a deadly bite. And when you consider that they're fiercely protective of their young, it's clear why hippos are not to be messed with. For such a seemingly peaceful animal, they sure pack a punch!

Q: What is the only mammal known to have scales? A: The pangolin.

Q: Which mammal has a complex digestive system with four stomach chambers to break down cellulose? A: The cow (a ruminant).

Q: What mammal species is known for having a prehensile tail, used to grasp and hold objects? A: The opossum.

Q: Which mammal can communicate using infrasound, sound waves below the range of human hearing? A: The elephant.

Q: Which mammal's kidneys are so efficient they can survive without ever drinking water? A: The kangaroo rat.

Q: What is the only mammal with true horns that are not shed annually? A: The pronghorn antelope.

Q: Which marine mammal has the thickest blubber layer? A: The bowhead whale.

Q: What mammal species has developed the largest social groups of any wild species, sometimes numbering in the millions? A: The African naked mole-rat.

Q: Which mammal is capable of lowering its heart rate to just a few beats per minute while diving? A: The sperm whale.

The sperm whale plays a legendary role in the real-life tragedy of the Essex, a whaling ship from Nantucket that was sunk by an enraged sperm whale in 1820. The Essex crew had been hunting whales for their oil—a valuable resource at the time—when a massive sperm whale, reportedly 85 feet long, rammed the ship twice, causing it to sink. The crew was left stranded at sea for months, facing starvation and extreme conditions.

This dramatic incident became the inspiration for Herman Melville's classic novel *Moby-Dick*. In the book, the white whale symbolizes the ultimate, untamed force of nature, with Captain Ahab obsessed with hunting it down. Sperm whales, the largest of the toothed whales, are powerful creatures, known for their massive heads and deep-diving abilities. The story of the Essex is a haunting reminder of how these giants of the sea can turn the tables on their hunters, leaving a lasting mark on history and literature.

Q: Which mammal is known for its specialized tusks used primarily for digging through ice and sensing the sea floor? A: The walrus.

Q: What mammal is the largest carnivore native to Antarctica? A: The leopard seal.

Q: Which mammal uses a technique called "spinning" to corral fish in hunting? A: The orca (killer whale).

Q: Which mammal species has the largest brain-to-body size ratio, second only to humans? A: The dolphin.

Q: Which nocturnal mammal uses a combination of taste and echolocation to find its way? A: The fruit bat.

Q: What mammal has the thickest skin of any land animal? A: The rhinoceros.

Q: Which mammal's stomach secretes a highly acidic enzyme called "rennet" used in curdling milk? A: The calf (baby cow).

Q: Which mammal is known for using unique vocal "signatures" to identify family members? A: The bottlenose dolphin.

Q: What mammal can sleep for up to 75% of its life in a state of torpor to conserve energy? A: The little brown bat.

Q: Which mammal has teeth with the same structure and composition as human teeth but grows continuously throughout its life? A: The beaver.

Q: What is the only mammal known to have a cloaca, an opening for both excretion and reproduction? A: The monotreme (e.g., platypus and echidna).

Q: Which mammal is known to have the largest group-living carnivore social structure? A: The African wild dog.

Q: What mammal holds the record for the highest recorded speed in the ocean? A: The common dolphin, reaching speeds of up to 60 km/h (37 mph).

Q: Which mammal exhibits the highest frequency of ultrasonic vocalization, used to communicate with its young? A: The greater spear-nosed bat.

Q: What mammal can change the shape of its eyes to focus underwater? A: The seal.

Q: Which mammal is known to produce milk that is blue in color due to high levels of casein and iron? A: The hooded seal.

Q: What mammal has the most flexible backbone, allowing for incredible agility and speed when hunting? A: The cheetah.

Q: Which mammal is known to "moonwalk" or shuffle backward to spread scent markers on trees? A: The maned wolf.

Q: Which mammal can recognize individual human faces and even remember them for years? A: The elephant.

Q: What mammal is capable of creating intricate nests made of leaves and branches high up in trees? A: The orangutan.

Dogs seem to have stomachs made of steel when it comes to eating things that would make humans seriously ill, like rotten or decomposing food. While it's not always safe for them, dogs have a few advantages that help them handle gross, bacteria-laden foods better than we can. For one, dogs have strong stomach acid, which helps break down food quickly and can kill off many harmful bacteria that might make humans sick. Their digestive systems are designed to handle raw meat and other things that humans would never touch.

Dogs are also descendants of wolves, who survive in the wild by scavenging and eating whatever they can find, including old, decaying meat. This scavenging instinct still runs strong in domestic dogs. Plus, they have a relatively short digestive tract, which means that harmful bacteria don't have as much time to take hold and cause infections.

That said, while dogs are more resistant to rotten food, it doesn't mean they're immune to it. Some bacteria and toxins can still harm them, and too much of the wrong thing could lead to food poisoning or even more serious conditions like pancreatitis. So, while they might handle it better than humans, it's still a good idea to keep them away from the trash!

Q: Which mammal uses a unique, ultraviolet-sensitive vision to detect urine trails left by prey? A: The reindeer.

Q: Which mammal has the longest period of parental care among carnivores? A: The polar bear, with cubs staying with their mother for up to 2.5 years.

Q: What mammal has the largest vocal range of any animal, capable of producing sounds at frequencies too high for humans to hear? A: The greater bulldog bat.

Q: Which mammal species has the highest concentration of capillaries in its muscles, allowing for extended deep dives? A: The Weddell seal.

Q: What mammal can alter the pitch of its vocalizations depending on the social context, a behavior linked to complex emotional states? A: The beluga whale.

The beluga whale might look cute with its smiling face and snowy white skin, but there's a funny legend surrounding them—they're rumored to have really bad breath! Belugas are toothed whales, and their diet consists mainly of fish, squid, and crustaceans, which can definitely lead to some fishy breath.

While it's mostly a joke, the truth is that belugas, like other marine animals, don't have the luxury of toothpaste or minty fresh mouthwash. In the wild, their breath might carry the strong scent of their last meal—imagine a seafood buffet that's been sitting out a bit too long! Since belugas are social animals, often hanging out in pods and communicating with clicks, whistles, and sounds, you can only hope their whale friends don't mind the smell.

Of course, bad breath aside, belugas are fascinating creatures known for their intelligence and playful behavior. But it's probably best to admire them from a distance!

Q: Which mammal produces the densest milk in terms of fat content, allowing its young to grow quickly in harsh environments? A: The blue whale.

Q: What mammal can lower its body temperature to near freezing during hibernation to conserve energy? A: The Arctic ground squirrel.

Q: Which mammal produces a unique "purring" sound when content, similar to domestic cats? A: The cheetah.

Q: What mammal is known for using its highly sensitive whiskers to detect vibrations in the water? A: The sea otter.

Q: Which mammal's tongue can be longer than its entire body length? A: The tube-lipped nectar bat.

Q: What mammal has the unique ability to twist its limbs in nearly any direction, aiding in its climbing agility? A: The margay (a small wild cat from South America).

Q: Which mammal's nostrils can close completely when diving underwater? A: The hippopotamus.

Q: What mammal has a specially adapted ankle joint that allows it to rotate its feet 180 degrees to climb down trees headfirst? A: The margay.

Are you ready to go Latin? If you dare, here are some of the more common mammals and their scientific names.

The following animal names follow what is known as the binomial nomenclature system, with the genus listed first (capitalized) and the species second (lowercase).

Homo sapiens – Human
Panthera leo – Lion
Panthera tigris – Tiger
Felis catus – Domestic Cat
Bos taurus – Domestic Cow
Elephas maximus – Asian Elephant
Loxodonta africana – African Elephant
Gorilla gorilla – Western Gorilla
Balaenoptera musculus – Blue Whale
Ursus arctos – Brown Bear
Equus ferus caballus – Domestic Horse
Vulpes vulpes – Red Fox
Phascolarctos cinereus – Koala

Ornithorhynchus anatinus – Platypus
Phoca vitulina – Harbor Seal
Odobenus rosmarus – Walrus
Capra aegagrus hircus – Domestic Goat
Sus scrofa domestica – Domestic Pig
Castor canadensis – North American Beaver
Ovis aries – Domestic Sheep
Macropus rufus – Red Kangaroo
Rangifer tarandus – Reindeer (Caribou)
Mustela putorius furo – Ferret
Pongo pygmaeus – Bornean Orangutan

The scientific name for the domestic dog is Canis lupus familiaris. It is a subspecies of the gray wolf, Canis lupus, and is part of the family Canidae.

If you're ever in Philadelphia and decide to grab a cheesesteak, remember this, there are only two words they want to hear at Pat's or Geno's, "wit" or "witout." That is, with onions or without onions. Similarly, while scientists typically use Linnaean Taxonomy to classify and order the animal kingdom, you can also describe animals as either vertebrate or invertebrate, meaning with or without a backbone.

More specifically, the main difference between vertebrates and invertebrates comes down to one thing: the backbone. Vertebrates are animals that have a spine or backbone, while invertebrates are those that don't. It's really that simple!

Vertebrates include familiar animals like mammals, birds, reptiles, amphibians, and fish. Think of your dog, a robin, or a lizard—these are all vertebrates. They have a solid internal skeleton made of bone or cartilage, which gives their bodies structure and helps them move around. Vertebrates also tend to have more complex body systems, like advanced nervous systems, making them better at responding to their environment.

Invertebrates, on the other hand, include animals like insects, spiders, jellyfish, and worms—basically, all the creatures without a backbone. In fact, most animals on Earth are invertebrates. These creatures usually have simpler body structures, but they're incredibly diverse. Some have hard exoskeletons, like crabs, while others, like jellyfish, are soft and squishy.

Invertebrates are often smaller and less complex than vertebrates, but they've got their own superpowers, like the ability to regrow body parts or survive extreme conditions. So, while vertebrates get a lot of attention, invertebrates are just as fascinating and important!

When you're ready to get creepy and check out the invertebrate side of life, turn the page and delve into the world of insects.

Insects

Insects are a type of small animal that make up the largest group of living creatures on Earth. They have a few key features that set them apart from other animals. All insects have a body divided into three parts: the head, thorax, and abdomen. They also have six legs attached to their thorax, and most insects have wings at some point in their life, though not all of them can fly.

Another important characteristic is that insects have a hard exoskeleton, which acts like armor to protect their bodies. Unlike humans, who have bones inside their bodies, insects wear their skeleton on the outside. Insects also have antennae on their heads, which they use to sense their surroundings, such as detecting smells, vibrations, or changes in temperature.

Insects go through a process called metamorphosis, which means they change form as they grow. Many insects, like butterflies, start as larvae (like caterpillars) before transforming into their adult form.

Insects are found almost everywhere on Earth, from forests to deserts, and they play important roles in nature, such as pollinating plants, decomposing dead matter, and serving as food for other animals. Some common examples of insects include ants, bees, butterflies, and beetles.

Q: Which insect is responsible for pollinating about one-third of the food we eat? A: The honeybee (Apis mellifera).

Q: What is the fastest flying insect in the world? A: The dragonfly, which can reach speeds of up to 35 mph.

Q: Which insect has ears located on its legs? A: The cricket.

Q: What is the largest insect by weight? A: The giant weta (Deinacrida spp.) from New Zealand.

Q: Which insect is known for producing silk? A: The silkworm moth (Bombyx mori).

Q: How many eyes does a common housefly have? A: Five eyes—two large compound eyes and three smaller simple eyes (ocelli).

Q: What is the longest-living insect? A: The queen termite, which can live up to 50 years.

Q: Which insect can walk on water due to its hydrophobic legs? A: The water strider (Gerridae family).

Q: What is the only insect that can turn its head from side to side? A: The praying mantis.

Q: Which insect migrates the longest distance? A: The monarch butterfly, migrating up to 3,000 miles.

Q: What insect is known to "play dead" when threatened? A: The rove beetle.

Q: What insect spends most of its life as a larva before emerging as an adult for only a few days? A: The mayfly.

Q: Which insect is famous for using "chemical warfare" to protect itself? A: The bombardier beetle, which sprays a hot chemical mixture from its abdomen.

Q: What insect can carry objects 50 times its own body weight? A: The ant.

Q: Which insect is known for mimicking the appearance of flowers to capture prey? A: The orchid mantis.

Q: What insect creates a buzzing sound by rubbing its wings together? A: The cicada.

Q: Which insect has the most extensive colonies, sometimes housing over a million individuals? A: The termite.

Termites may be tiny, but they're absolute nightmares for homeowners! These sneaky insects are often called "silent destroyers" because they can chew through wood, flooring, and even wallpaper—basically anything with cellulose—and you might not even realize they're there until the damage is done. A single termite colony can have millions of termites, and they can eat through the wooden structures of a house 24/7.

For homeowners, this means serious trouble. Termite damage can weaken the foundation of your home, cause floors to sag, and even lead to costly repairs—sometimes running into the thousands of dollars. Worst part? Insurance doesn't usually cover termite damage because it's considered preventable. Yikes!

Termites thrive in warm, humid climates, but they can be a problem almost anywhere. Regular inspections and preventative treatments are key if you want to avoid sharing your home with these hungry pests. Otherwise, they'll eat your house—literally—from the inside out!

Q: What insect uses the sun to navigate long distances? A: The desert ant (Cataglyphis).

Q: Which insect has the ability to regenerate lost limbs? A: The stick insect.

Q: Which insect's larvae are known as "armyworms" because of the way they march in groups? A: The moth caterpillar.

Q: What is the most poisonous insect in the world? A: The harvester ant, which has venom potent enough to kill small animals.

Q: Which insect is known for living symbiotically with certain species of ants? A: The aphid.

Q: Which insect lays its eggs in fresh cow dung? A: The dung beetle.

Q: What insect is known for its bioluminescent abilities? A: The firefly.

Q: Which insect can survive freezing temperatures by producing an antifreeze-like compound? A: The woolly bear caterpillar.

Q: What insect is responsible for producing honey? A: The honeybee.

Honeybees might seem small, but they play a huge role in food production. These buzzing little workers are responsible for pollinating a massive portion of the crops we rely on. In fact, about one-third of the food we eat depends on pollinators like honeybees. Without them, many of our favorite foods—fruits, vegetables, nuts—would be in short supply.

As honeybees fly from flower to flower collecting nectar, they accidentally transfer pollen, helping plants reproduce. This process is vital for crops like apples, almonds, berries, cucumbers, and many more. Without bees, farmers would have a hard time growing these crops in abundance.

Q: Which insect can move at speeds of up to 120 body lengths per second? A: The Australian tiger beetle.

Q: What insect builds hives out of paper it makes from chewed wood fibers? A: The wasp.

Q: Which insect species has the highest frequency of wing beats per second? A: The midge, with up to 1,000 beats per second.

Q: What insect uses a "dance" to communicate the location of food to other members of its hive? A: The honeybee.

Q: Which insect is known for its ability to fly backward, sideways, and hover? A: The hoverfly.

Q: What insect undergoes complete metamorphosis, changing from egg to larva to pupa to adult? A: The butterfly.

Q: Which insect has a specialized proboscis for drinking nectar from flowers? A: The butterfly.

Q: What insect is known for its bright, iridescent wings and delicate flight? A: The dragonfly.

Q: Which insect can leap distances over 100 times its body length? A: The flea.

Q: What insect has the largest wingspan of any living insect? A: The atlas moth, with a wingspan of up to 12 inches.

Q: Which insect spends up to 17 years underground before emerging as an adult? A: The periodical cicada.

Q: What insect can see ultraviolet light, which helps it locate flowers? A: The bee.

Q: Which insect uses chemical signals called pheromones to communicate with others of its species? A: The ant.

Q: What insect is known for mimicking dead leaves as camouflage? A: The leaf insect.

Q: Which insect has the strongest bite force relative to its size? A: The trap-jaw ant.

Q: What insect can survive without its head for up to a week? A: The cockroach.

Q: Which insect larvae are known for building silk shelters or "tents" in trees? A: The tent caterpillar.

Q: What insect has a built-in "jet propulsion" system, using water pressure to launch itself into the air? A: The water boatman.

Q: Which insect is capable of parthenogenesis, meaning it can reproduce without mating? A: The aphid.

Q: What insect is known for its highly developed social structure, including queens, workers, and drones? A: The honeybee.

Q: Which insect is capable of producing both winged and wingless offspring depending on environmental conditions? A: The aphid.

Q: What insect is famous for "bleeding" a chemical defense fluid from its joints when threatened? A: The ladybug.

Q: Which insect uses an air bubble on its abdomen to breathe while underwater? A: The diving beetle.

Q: What insect has a powerful ovipositor used to deposit eggs into the bodies of other insects? A: The ichneumon wasp.

Q: Which insect can be found at altitudes as high as 6,000 meters (19,685 feet) in the Himalayas? A: The bumblebee.

Q: What insect has wings covered with microscopic scales that reflect light? A: The butterfly.

Q: Which insect is known for engaging in "tug-of-war" behavior over prey? A: The antlion larva.

Q: What insect can remain in a state of diapause (suspended development) for several years to survive harsh conditions? A: The silk moth.

Q: Which insect can spray a noxious chemical to deter predators? A: The stink bug.

Q: What insect is the smallest known, measuring less than 0.5 mm in length? A: The fairyfly (a type of parasitic wasp).

Q: Which insect uses iridescent scales on its wings for camouflage and signaling? A: The butterfly.

Q: What insect is known for its precision "swarm" behavior during migration? A: The locust.

Q: Which insect larvae are known to spin cocoons made of silk?
A: The moth caterpillar.

Q: What insect's wings create a high-pitched "buzzing" sound as they vibrate? A: The mosquito.

Q: Which insect has specially adapted mouthparts for piercing and sucking blood? A: The mosquito.

Mosquitoes might just be the most annoying creatures on the planet, but they're more than just a nuisance—they're also deadly disease vectors. A "vector" means they can carry and spread diseases to humans without getting sick themselves. Some of the most dangerous diseases mosquitoes spread include malaria, dengue fever, Zika virus, and West Nile virus.

These little bloodsuckers pick up the disease-causing pathogens when they bite an infected person or animal. Then, when they bite someone else, they inject the pathogens into the next person's bloodstream, spreading the disease. Malaria alone causes hundreds of thousands of deaths every year, mainly in tropical and subtropical regions.

Mosquitoes thrive in warm, humid areas, and stagnant water is their breeding ground, which makes controlling them tricky. In places where mosquito-borne diseases are common, preventative measures like mosquito nets, repellents, and eliminating standing water are key to staying safe. These tiny pests are a big health threat, making them one of the deadliest animals to humans!

Q: What insect produces a secretion known as "royal jelly" to feed the queen? A: The honeybee.

Q: Which insect's larvae are often called "white grubs" and are known to damage plant roots? A: The June beetle.

Q: What insect uses sound to communicate by rubbing body parts together in a process called "stridulation"? A: The cricket.

Q: Which insect larvae are known as "wigglers" because of their rapid swimming movements? A: Mosquito larvae.

Q: What insect has a unique "umbrella" structure in its wings to help trap air and float on water? A: The water spider.

Q: Which insect feeds on nectar and pollen and is a vital pollinator for many crops? A: The bumblebee.

Q: What insect's nymphs are called "instars" because they molt multiple times before becoming adults? A: The grasshopper.

Q: Which insect is capable of producing an extremely loud chirping noise during mating season? A: The cicada.

Q: What insect has mandibles strong enough to slice through leaves and wood? A: The leafcutter ant.

Q: Which insect secretes wax to build its hive? A: The honeybee.

Q: What insect species creates mud tunnels to protect itself and its food from the elements? A: The termite.

Q: Which insect has the ability to "see" polarized light, allowing it to navigate more effectively? A: The dragonfly.

Q: What insect can survive being submerged in water for over 24 hours by trapping air bubbles around its body? A: The diving bell spider.

Q: Which insect is capable of changing its body color to blend in with its surroundings? A: The walking stick insect.

Q: What insect's larvae produce a sticky substance used to defend against predators? A: The antlion larva.

Q: Which insect uses a form of chemical "alarm" pheromone to signal danger to its colony? A: The ant.

Q: What insect is known for engaging in ritualized combat, often "locking" mandibles with its opponent? A: The stag beetle.

Q: Which insect is known for its parasitic behavior, laying its eggs in the nests of other insects? A: The cuckoo bee.

Q: What insect undergoes incomplete metamorphosis, meaning it hatches as a miniature version of the adult? A: The grasshopper.

Q: Which insect larvae are called "maggots" and are known for their role in decomposing dead animals? A: The fly.

Q: What insect is responsible for producing the natural dye cochineal, used in food coloring? A: The cochineal scale insect.

Q: Which insect has specialized scales on its wings that allow for rapid flight changes? A: The butterfly.

Q: What insect has specialized jaws capable of delivering a venomous bite? A: The centipede.

Q: Which insect is known for constructing "burrow" nests in the ground? A: The cicada killer wasp.

Q: What insect feeds on the blood of mammals, birds, and reptiles, often spreading diseases? A: The tick (arachnid).

Q: Which insect species is known for "dancing" in mid-air during courtship rituals? A: The gnat.

Q: What insect is commonly known as a "winged aphid" when it develops wings in certain generations? A: The aphid.

Q: Which insect is capable of producing a "smoke screen" of scales to evade predators? A: The moth.

Q: What insect uses vibrations from sound waves in the air to locate mates? A: The mosquito.

Q: Which insect's larvae are known as "cutworms" because they cut off young plants at the soil level? A: The moth.

Q: What insect's wings are covered with tiny hairs that help it sense changes in air pressure? A: The moth.

Q: Which insect is known to perform a unique "flight display" to attract mates? A: The mayfly.

Q: What insect can leap over 100 times its body length using its powerful hind legs? A: The grasshopper.

At first glance, grasshoppers and locusts look pretty much the same, but there's a big difference in their behavior. Grasshoppers are solitary creatures that hop around, minding their own business, munching on plants. They're harmless in small numbers and don't usually cause a fuss.

Locusts, on the other hand, are what grasshoppers turn into when conditions are just right—or wrong, depending on how you look at it. When food is abundant and there's a lot of moisture, grasshoppers can morph into locusts, becoming more social and gathering into massive swarms. These swarms can be devastating, wiping out crops and causing food shortages, especially in parts of Africa and Asia.

So, technically, a locust is just a grasshopper with a bad attitude when it's part of a swarm. The environmental triggers cause a change in their behavior and even their appearance, leading to one of nature's most destructive insect transformations!

Q: Which insect can store pollen in special "baskets" on its legs for transport? A: The honeybee.

Q: What insect uses a specialized mouthpart called a "proboscis" to feed on liquid? A: The mosquito.

Q: Which insect's larvae are known for spinning "leaf rolls" in which to hide and feed? A: The leafroller moth.

Q: What insect is capable of producing ultrasonic clicks to deter bats? A: The tiger moth.

Q: Which insect is responsible for producing ambergris, a substance used in perfumes? A: The sperm whale (actually a mammal, not an insect — but worth a mention!).

Q: What insect's wings are shaped like "teardrops," helping it glide through the air? A: The damselfly.

Advanced Fun Facts

Q: Which insect has the largest number of individual muscles, with over 900? A: The caterpillar.

Q: What insect uses a process called "trophallaxis" to share food with colony members? A: The ant.

Q: Which insect is known to lay its eggs inside the living body of another insect, which will eventually be consumed from the inside? A: The parasitic wasp.

Q: What insect can fly in a straight line for up to 500 miles without stopping? A: The desert locust.

Q: Which insect is capable of producing sound by rubbing its wings together, a process called "crepitation"? A: The grasshopper.

Q: What insect can be found living in geothermal springs with water temperatures above 50°C (122°F)? A: The thermophilic insect, Ephydra bruesi.

Q: Which insect uses a specialized enzyme called luciferase to produce light? A: The firefly.

Q: What insect's larvae are often mistaken for worms because of their cylindrical, legless bodies? A: The crane fly.

Q: Which insect can detect pheromones over a distance of several miles using its antennae? A: The male moth.

Q: What insect creates "galls," or growths, on plants where it deposits its eggs? A: The gall wasp.

Q: Which insect is known to form "living bridges" using their own bodies to help other members cross gaps? A: The army ant.

Q: What insect is known to have a heat tolerance of up to 122°F (50°C)? A: The Sahara desert ant.

Q: Which insect has larvae that can go through diapause, suspending development for several years in response to environmental conditions? A: The gypsy moth.

Q: What insect's exoskeleton contains chitin, a molecule that is also found in the cell walls of fungi? A: All arthropods, including insects.

Q: Which insect is known for its ability to mimic the appearance of bird droppings to avoid predators? A: The swallowtail caterpillar.

Q: What insect has a life cycle in which adults do not eat, but solely focus on reproduction before dying? A: The mayfly.

Q: Which insect is capable of producing "honeydew," a sugary liquid consumed by ants? A: The aphid.

Q: What insect has mouthparts designed specifically for sponging up liquids? A: The housefly.

Q: Which insect has been found to produce antibiotics to protect itself from harmful bacteria and fungi? A: The leafcutter ant, via its symbiotic relationship with a fungus.

Q: What insect exhibits a phenomenon called "telescoping generations," where multiple generations are developing inside each other at the same time? A: The aphid.

Q: Which insect has been shown to be able to detect infrared radiation, helping it locate warm-blooded prey? A: The pyrophilic beetle (Melanophila acuminata).

Q: What insect's larvae are known for their "doodlebug" behavior, where they create cone-shaped traps in sandy soil? A: The antlion.

Q: Which insect can change its body color in response to temperature or humidity? A: The desert locust.

Q: What insect has the ability to create an air bubble around its body to breathe underwater? A: The backswimmer.

Q: Which insect uses a special "vibration dance" to communicate potential food sources? A: The honeybee, through its "waggle dance."

Q: What insect produces silk from its mouth to create cocoons or shelters? A: The caddisfly.

Q: Which insect can see a wider spectrum of colors than humans, including ultraviolet light? A: The mantis shrimp (though technically a crustacean, its visual abilities are insect-like).

Q: What insect's males can store sperm in a special organ, allowing them to reproduce for months after mating? A: The queen ant.

Q: Which insect is capable of "subsocial" behavior, where parents provide care to their offspring after they hatch? A: The burying beetle.

Q: What insect exhibits one of the longest periods of aerial gliding? A: The planthopper nymph.

Q: Which insect has the highest known density of hair-like structures called "setae" on its body, used for sensing its surroundings? A: The tarantula hawk wasp.

Q: What insect has larvae known as "hellgrammites" and are used as bait for fishing? A: The dobsonfly.

Q: Which insect is known for "trail-following" behavior, using pheromones to lead others to food? A: The ant.

Q: What insect has its spiracles located on the abdomen, allowing it to breathe while buried in soil? A: The cicada nymph.

Q: Which insect is capable of detecting sound through its legs and using it for mating calls? A: The katydid.

Q: What insect has specialized "feathered" wings designed to attract mates and also for flight? A: The plume moth.

Q: Which insect can detect changes in the Earth's magnetic field to navigate during migration? A: The monarch butterfly.

Q: What insect has males that produce an acoustic "love song" by rubbing their wings together to attract mates? A: The bush-cricket.

Q: Which insect is known for using mimicry to resemble ants in order to avoid predators? A: The myrmecomorphic spider (though technically an arachnid, many insects use this tactic too).

Cockroaches are definitely insects, and they're part of the order Blattodea. While they get a bad rap (and for good reason), cockroaches have been around for millions of years, adapting to just about every environment on Earth. Yes, they're notorious for invading homes and are known to be vectors for diseases. Cockroaches can carry bacteria like Salmonella and E. coli on their bodies, which can contaminate food and surfaces. They've also been linked to triggering allergies and asthma, especially in sensitive individuals, thanks to the allergens in their droppings, saliva, and shed skins.

But are they all bad? Believe it or not, cockroaches do serve a purpose in the ecosystem. In the wild, they act as decomposers, breaking down dead plant material, decaying matter, and even animal waste. This helps recycle nutrients back into the soil, supporting plant growth and overall ecosystem health. So, while they're definitely not welcome guests in our kitchens, they're pretty important when it comes to cleaning up organic waste in the wild.

Despite their creepy reputation, cockroaches play a role in maintaining balance in nature—though we'd all agree they're best kept outside the house!

Q: What insect species is capable of "rafting," where individuals link together to form floating structures during floods? A: The fire ant.

Q: Which insect has compound eyes with over 30,000 lenses, providing almost 360-degree vision? A: The dragonfly.

Q: What insect uses its proboscis to inject digestive enzymes into its prey before consuming it? A: The assassin bug.

Q: Which insect produces a sticky secretion from its feet to allow it to walk upside down on smooth surfaces? A: The fly.

Q: What insect can detect vibrations in the ground caused by other insects, allowing it to ambush them? A: The mole cricket.

Q: Which insect has larvae that form protective cases from sand, plant matter, and silk? A: The caddisfly.

Q: What insect produces chemicals that act as tranquilizers, paralyzing its prey before feeding? A: The spider wasp.

Here are some of the more common insects and their scientific names.

Apis mellifera – Western Honeybee
Danaus plexippus – Monarch Butterfly
Anopheles gambiae – African Malaria Mosquito
Drosophila melanogaster – Common Fruit Fly
Atta cephalotes – Leafcutter Ant
Locusta migratoria – Migratory Locust
Musca domestica – Housefly
Cimex lectularius – Common Bedbug
Gryllus campestris – Field Cricket
Acheta domesticus – House Cricket
Aedes aegypti – Yellow Fever Mosquito
Coccinella septempunctata – Seven-Spotted Ladybug
Bombyx mori – Silkworm Moth
Formica rufa – Red Wood Ant
Tenebrio molitor – Mealworm Beetle
Lepisma saccharina – Silverfish
Periplaneta americana – American Cockroach
Phyllophaga spp. – June Beetle
Pterophylla camellifolia – True Katydid
Manduca sexta – Carolina Sphinx Moth

Arachnids

Arachnids are a group of animals that include spiders, scorpions, ticks, and mites. They are different from insects in a few important ways. First, arachnids have eight legs, while insects only have six. Their bodies are also divided into two main parts: the cephalothorax (which is a combination of the head and thorax) and the abdomen.

Unlike insects, arachnids don't have antennae. Instead, they rely on other senses, like touch, to understand their environment. Many arachnids, especially spiders, have special fangs or pincers, which they use to catch and eat their prey. Spiders, for example, often produce silk to spin webs that trap insects for food.

Most arachnids are carnivorous, meaning they eat other animals, mostly insects or small creatures. Some, like ticks and mites, are parasites, meaning they live on or inside other animals, feeding off them.

Arachnids are usually found on land, in many different environments, from forests to deserts. They play important roles in ecosystems by helping control insect populations. Even though some people fear arachnids like spiders or scorpions, most are harmless to humans.

Q: What is the largest species of spider in the world by leg span? A: The giant huntsman spider.

Q: How many legs do all arachnids have? A: Eight legs.

Q: What arachnid can survive without food for over a year? A: The scorpion.

Q: Which arachnid has the ability to regenerate lost legs? A: The harvestman (also known as daddy longlegs).

Q: Which spider can trap bubbles of air on its abdomen to breathe underwater? A: The diving bell spider.

Q: What is the most venomous spider in the world? A: The Brazilian wandering spider.

Q: Which arachnid produces silk that is five times stronger than steel by weight? A: The orb-weaver spider.

Q: How do tarantulas defend themselves from predators? A: By flicking urticating (irritating) hairs from their abdomen.

Q: Which arachnid can paralyze its prey by spitting a sticky substance? A: The spitting spider.

Q: What arachnid is capable of "ballooning" by releasing silk threads to travel on the wind? A: The spiderling (young spiders).

Q: Which arachnid has claws similar to those of crabs or lobsters? A: The scorpion.

Q: What arachnid has no venom glands and instead relies on strong jaws to capture prey? A: The camel spider (solifuge).

Q: Which spider species exhibits sexual cannibalism, where the female eats the male after mating? A: The black widow spider.

Q: What arachnid is known for using its long legs to "lasso" its prey? A: The harvestman.

Q: Which arachnid can "play dead" when threatened by a predator? A: The wolf spider.

Q: What species of arachnid can deliver a fatal sting to humans with its tail? A: The deathstalker scorpion.

Q: Which arachnid has specialized pedipalps that act like pincers for grasping prey? A: The whip scorpion.

Q: What arachnid builds its web in a spiral shape to catch insects? A: The orb-weaver spider.

Q: Which arachnid is known for mimicking ants as a form of camouflage? A: The ant-mimicking jumping spider.

Q: What spider is capable of jumping up to 50 times its own body length? A: The jumping spider.

Q: Which arachnid has the ability to glow under ultraviolet (UV) light? A: The scorpion.

Q: What arachnid produces a silk web but does not use it to catch prey? A: The trapdoor spider.

Q: Which arachnid uses its front legs like antennae to feel its surroundings? A: The whip spider (tailless whip scorpion).

Q: What arachnid can walk on water due to the hydrophobic properties of its legs? A: The fishing spider.

Q: Which arachnid is known for creating webs in a funnel shape? A: The funnel-web spider.

Q: What arachnid uses pheromones to communicate with other members of its species? A: The tarantula.

Q: Which arachnid has the ability to liquefy its prey before consuming it? A: The wolf spider.

Q: What arachnid is capable of delivering a venomous bite that can cause necrosis in humans? A: The brown recluse spider.

Q: Which arachnid uses silk to wrap and store its prey for later consumption? A: The orb-weaver spider.

Q: What species of spider can change its color to blend in with its environment? A: The crab spider.

Q: Which arachnid is known for engaging in elaborate courtship dances? A: The peacock spider.

Q: What arachnid has specialized sensory hairs to detect vibrations in the air? A: The tarantula.

Q: Which spider can survive freezing temperatures by producing antifreeze proteins? A: The arctic wolf spider.

Q: What arachnid is commonly mistaken for a spider but is not a true spider? A: The harvestman.

Q: Which spider species carries its egg sac attached to its body? A: The wolf spider.

Q: What arachnid creates a "silk snare" to capture its prey rather than using a web? A: The bolas spider.

Q: Which spider can live for up to 30 years in captivity? A: The Mexican redknee tarantula.

Q: What arachnid is capable of "molting" its exoskeleton as it grows? A: All arachnids.

Q: Which arachnid can deliver a venomous bite that is medically significant to humans but rarely fatal? A: The black widow spider.

Q: What arachnid has evolved to hunt by running down its prey instead of using a web? A: The wolf spider.

Q: Which arachnid has evolved to "live in harmony" with ants, using them for protection? A: The ant-mimicking spider.

Q: What arachnid is considered one of the smallest species of spider in the world? A: The Patu digua spider.

Q: Which arachnid has venom powerful enough to paralyze insects but is generally harmless to humans? A: The jumping spider.

Q: What arachnid is known for building its web in dark, hidden places? A: The cellar spider.

Q: Which arachnid is known for creating a "burrow" lined with silk in the ground? A: The trapdoor spider.

Q: What species of scorpion is known for its unusually long tail? A: The emperor scorpion.

Q: Which arachnid has been observed "team-hunting" in groups? A: The social spider (Stegodyphus spp.).

Q: What arachnid can create a "decoy" version of itself out of silk to confuse predators? A: The Cyclosa spider.

Q: Which arachnid has evolved to live on the surface of water? A: The fishing spider.

Q: What arachnid species has been found to produce the loudest sound made by any spider? A: The stridulating tarantula.

Q: Which arachnid can remain completely still for hours as a form of ambush hunting? A: The crab spider.

Q: What arachnid species is the most widespread globally? A: The house spider.

Q: Which arachnid uses "trap" doors to ambush its prey? A: The trapdoor spider.

Q: What arachnid can hunt and catch small fish with its powerful front legs? A: The fishing spider.

Q: Which arachnid can regenerate missing eyes during molting? A: Some species of tarantulas.

Q: What arachnid is known to dig elaborate tunnels underground? A: The trapdoor spider.

Q: Which arachnid is the fastest-moving spider, clocking in at speeds of up to 0.5 meters per second? A: The giant house spider.

Q: What arachnid uses vibrations in its web to detect prey? A: The orb-weaver spider.

Q: Which spider species is capable of carrying up to 100 babies on its back after hatching? A: The wolf spider.

Q: What species of scorpion is known for having a venomous sting that can cause extreme pain but is rarely fatal? A: The bark scorpion.

Q: Which arachnid species is the most venomous in North America? A: The Arizona bark scorpion.

The Arizona bark scorpion might be small, but it's the most dangerous scorpion in North America. These little guys are only about 3 inches long, but their sting packs a punch! Found in the southwestern U.S., particularly in Arizona, they love to hide in cool, dark places like woodpiles, shoes, or even inside homes.

What makes the Arizona bark scorpion a serious concern is its potent venom. While most scorpion stings are painful but not life-threatening, a sting from this scorpion can cause intense pain, numbness, muscle twitching, and in some cases, breathing difficulties. For small children, the elderly, or those allergic to the venom, it can even be life-threatening without medical treatment.

If you live in an area with bark scorpions, shaking out shoes and being cautious around dark, damp spots is a must. While stings are rarely fatal, they're definitely not something you want to experience firsthand!

Q: What arachnid has specialized "spinning" glands to produce different types of silk? A: All spiders.

Q: Which spider can detect the polarization of light and use it for navigation? A: The jumping spider.

Q: What arachnid is capable of living its entire life in caves without ever seeing sunlight? A: The cave spider.

Q: Which spider species has a unique "social structure" where several individuals live together in one web? A: The social spider.

Q: What arachnid can capture insects by mimicking the smell of a female moth to attract male moths? A: The bolas spider.

Q: Which arachnid is known to wrap its prey in silk before consuming it? A: The tarantula.

Q: What species of arachnid is known for its ability to create spiral-shaped webs to trap insects? A: The orb-weaver spider.

Q: Which arachnid can trap prey using a "net" made of silk? A: The ogre-faced spider.

Q: What arachnid is known for storing sperm in a specialized sac until it is needed for reproduction? A: The tarantula.

Q: Which spider has one of the fastest attack speeds in the animal kingdom? A: The trap-jaw spider.

Q: What arachnid is known for digging a shallow pit in the sand to trap ants and other insects? A: The antlion (though not a true arachnid, it's often confused as one).

Q: Which arachnid has evolved to live inside human homes without needing to go outside? A: The common house spider.

Q: What species of scorpion has venom that contains compounds being studied for medical use? A: The deathstalker scorpion.

Q: Which arachnid can produce a web that is nearly invisible to the naked eye? A: The golden orb-weaver spider.

Q: What arachnid uses hydraulic pressure to move its legs rather than muscles? A: The tarantula.

Q: Which arachnid has evolved to mimic bird droppings as a form of camouflage? A: The bird-dropping spider.

Q: What arachnid uses vibrations from its web to "measure" the size and strength of its prey? A: The orb-weaver spider.

Q: Which arachnid is known to "recycle" its web by eating it and using the proteins to build a new one? A: The garden spider.

Q: What arachnid is capable of jumping long distances despite having no muscles in its legs? A: The jumping spider.

Q: Which spider produces a silk "dragline" as it moves to catch itself if it falls? A: All jumping spiders.

Q: What arachnid uses "sexual cannibalism" as a form of mate selection, where the female often eats the male? A: The praying mantis (technically an insect but often associated with arachnid behavior).

Q: Which arachnid produces a venom that has no known antidote in certain species? A: The brown recluse spider.

Q: What arachnid can create silk that is visible under ultraviolet light? A: The orb-weaver spider.

Q: Which spider is capable of creating its web between two surfaces over six feet apart? A: The garden spider.

Q: What arachnid can survive in arid environments by conserving moisture in its exoskeleton? A: The desert scorpion.

Q: Which arachnid is known for burrowing in sand dunes and desert environments? A: The sand spider.

Q: What arachnid species is known to form intricate social hierarchies within colonies? A: The social spider.

Q: Which spider has silk that can stretch up to five times its original length without breaking? A: The Darwin's bark spider.

Q: What arachnid can survive months without food by slowing down its metabolism? A: The tarantula.

Q: Which arachnid has evolved to tolerate high levels of radiation? A: The common house spider.

Q: What arachnid can adapt its web-building techniques based on the availability of prey? A: The orb-weaver spider.

Q: Which spider has been known to "play dead" as a defense mechanism? A: The wolf spider.

Q: What arachnid produces venom that specifically targets the nervous system of its prey? A: The Brazilian wandering spider.

Talk about arachnophobia, get this…

The Brazilian wandering spider (Phoneutria), also known as the armed spider or banana spider, is one of the most venomous spiders in the world. Native to tropical South America, these spiders are known for their aggressive nature and their tendency to wander the forest floor at night in search of prey, rather than building webs. They are often found in homes, clothing, or even bananas, giving them their common name.

The venom of the Brazilian wandering spider contains a potent neurotoxin, which can cause muscle paralysis, extreme pain, and in rare cases, death, especially if untreated. Despite this, human fatalities are uncommon due to the availability of antivenom. Interestingly, the venom also causes an unusual side effect in males: prolonged, painful erections, which has led to research into its potential medical uses for erectile dysfunction.

Despite being highly dangerous, these spiders play an important role in their ecosystems, controlling insect populations and maintaining the balance of tropical environments.

Fine. If they agree to limit their activities to hunting insects, I'll promise to stay out of the Amazon jungle.

Q: Which arachnid has been found to have "learned" behaviors, such as improving hunting tactics? A: The jumping spider.

Q: What arachnid's venom is being studied for use as a non-addictive painkiller? A: The Chilean rose tarantula.

Q: Which arachnid has been found to have individual personalities, such as being more aggressive or timid? A: The tarantula.

Q: What arachnid is capable of building webs in areas with heavy foot traffic and vibrations without being disturbed? A: The common house spider.

Q: Which arachnid can manipulate its prey into immobilizing itself in the web? A: The orb-weaver spider.

Q: What arachnid species has evolved to hunt exclusively at night, relying on highly sensitive vision? A: The wolf spider.

Advanced Fun Facts

Q: Which arachnid has the largest venom glands in proportion to its body size? A: The Australian funnel-web spider.

The Australian funnel-web spider, particularly the Sydney funnel-web spider (Atrax robustus), is one of the most dangerous spiders in the world due to its potent venom and aggressive nature. Native to eastern Australia, these spiders are known for their distinctive funnel-shaped webs and burrows, which they use as a base for ambushing prey.

The venom of the funnel-web spider contains a neurotoxin called atraxotoxin, which can be lethal to humans, particularly young children. Bites from males are considered more dangerous as they tend to wander during mating season and have more potent venom. Symptoms of envenomation include severe pain, muscle spasms, difficulty breathing, and in extreme cases, death if left untreated.

Thankfully, an antivenom was developed in 1981, and since then, there have been no recorded deaths from funnel-web spider bites.

Q: What is the longest-lived arachnid species in captivity, reaching over 40 years? A: The Mexican redknee tarantula.

Q: Which spider species can vibrate its web to mimic the movement of wind to evade predators? A: The orb-weaver spider.

Q: What species of arachnid is known to produce a "safety line" made of silk when hunting? A: The jumping spider.

Q: Which arachnid can remain submerged in water for over 30 minutes without drowning? A: The raft spider.

Q: What arachnid is known to carry its egg sac in its jaws until the eggs hatch? A: The nursery web spider.

Q: Which arachnid can produce different types of silk for various purposes such as building webs, wrapping prey, and reproduction? A: The orb-weaver spider.

Q: What species of scorpion has evolved to survive in the harsh, cold environment of the Andes Mountains? A: The Andean scorpion.

Q: Which spider is known for its speed, reaching up to 10 mph while hunting? A: The camel spider (solifuge).

Q: What species of arachnid uses "mimicry" to resemble ants for protection against predators? A: The ant-mimicking jumping spider.

Q: Which arachnid has specialized hairs called trichobothria that detect air movement and vibrations? A: The tarantula.

While tarantulas have a fearsome reputation, their danger to humans is generally low. Tarantula bites are rare, and most species' venom is not particularly harmful to humans. A bite from a tarantula may cause mild pain, swelling, and itching, similar to a bee sting, but serious medical issues are extremely uncommon. Only a few species, such as the Indian ornamental tarantula (Poecilotheria regalis), have venom that can cause more intense symptoms, including muscle cramps or sweating, but still not life-threatening.

Tarantulas rely on their venom to subdue prey like insects, small rodents, and birds, but they are not aggressive toward humans unless provoked. More often, they rely on their urticating hairs—barbed bristles that can irritate the skin and eyes—as a defense mechanism. Overall, while intimidating, tarantulas pose little real danger to humans and are important in controlling insect populations in their environments.

Q: What is the primary purpose of the male tarantula's tibial spurs? A: To hold the female during mating and avoid being eaten.

Q: Which arachnid can "paralyze" its prey using a venomous bite and store it alive for later consumption? A: The tarantula hawk wasp (preys on tarantulas).

Q: What spider species builds a web with silk that reflects ultraviolet light to attract insects? A: The golden orb-weaver spider.

Q: Which arachnid has been observed displaying "sibling cooperation," where spiderlings help each other capture prey? A: The social spider (Stegodyphus spp.).

Q: What arachnid has evolved to produce "silk balloons" that allow it to float across water and colonize new areas? A: The fishing spider.

Q: Which arachnid uses silk threads to create "tripwires" outside its burrow to detect passing prey? A: The trapdoor spider.

Q: What arachnid can catch prey up to 15 times its own weight using its powerful legs and fangs? A: The wolf spider.

Q: Which spider is capable of building a multi-layered web with both sticky and non-sticky silk? A: The cobweb spider.

Q: What species of spider can launch itself at prey from a distance by using hydraulic pressure in its legs? A: The trap-jaw spider.

Q: Which spider is known for its speed, reaching up to 10 mph while hunting? A: The camel spider (solifuge).

Q: What species of arachnid uses "mimicry" to resemble ants for protection against predators? A: The ant-mimicking jumping spider.

Q: Which arachnid has specialized hairs called trichobothria that detect air movement and vibrations? A: The tarantula.

While tarantulas have a fearsome reputation, their danger to humans is generally low. Tarantula bites are rare, and most species' venom is not particularly harmful to humans. A bite from a tarantula may cause mild pain, swelling, and itching, similar to a bee sting, but serious medical issues are extremely uncommon. Only a few species, such as the Indian ornamental tarantula (Poecilotheria regalis), have venom that can cause more intense symptoms, including muscle cramps or sweating, but still not life-threatening.

Tarantulas rely on their venom to subdue prey like insects, small rodents, and birds, but they are not aggressive toward humans unless provoked. More often, they rely on their urticating hairs—barbed bristles that can irritate the skin and eyes—as a defense mechanism. Overall, while intimidating, tarantulas pose little real danger to humans and are important in controlling insect populations in their environments.

Q: What is the primary purpose of the male tarantula's tibial spurs? A: To hold the female during mating and avoid being eaten.

Q: Which arachnid can "paralyze" its prey using a venomous bite and store it alive for later consumption? A: The tarantula hawk wasp (preys on tarantulas).

Q: What spider species builds a web with silk that reflects ultraviolet light to attract insects? A: The golden orb-weaver spider.

Q: Which arachnid has been observed displaying "sibling cooperation," where spiderlings help each other capture prey? A: The social spider (Stegodyphus spp.).

Q: What arachnid has evolved to produce "silk balloons" that allow it to float across water and colonize new areas? A: The fishing spider.

Q: Which arachnid uses silk threads to create "tripwires" outside its burrow to detect passing prey? A: The trapdoor spider.

Q: What arachnid can catch prey up to 15 times its own weight using its powerful legs and fangs? A: The wolf spider.

Q: Which spider is capable of building a multi-layered web with both sticky and non-sticky silk? A: The cobweb spider.

Q: What species of spider can launch itself at prey from a distance by using hydraulic pressure in its legs? A: The trap-jaw spider.

Q: Which arachnid uses its "prey-pulling" method to drag insects into its burrow using strong chelicerae? A: The whip scorpion.

Q: What species of arachnid has a mating ritual where the male presents a gift of food to the female before mating? A: The nursery web spider.

Q: Which spider is known for living communally in a shared web, where up to hundreds of individuals live together? A: The communal orb-weaver spider.

Q: What arachnid uses specialized "spinnerets" to produce extremely fine silk for building egg sacs? A: The garden spider.

Q: Which arachnid can withstand up to 10 times the amount of radiation lethal to humans? A: The common house spider.

Q: What spider species can create an "elastic" web that stretches to capture larger prey? A: The Darwin's bark spider.

Q: Which arachnid relies on its pedipalps to taste and smell its surroundings? A: The scorpion.

Q: What arachnid has specialized legs that allow it to swim and hunt underwater? A: The water spider (Argyroneta aquatica).

Q: Which arachnid can detect changes in humidity through specialized sensory hairs? A: The cellar spider.

Q: What arachnid has the ability to digest its prey externally before consuming it as liquid? A: The tarantula.

Q: Which arachnid has evolved to produce venom that targets specific neurotransmitters in its prey? A: The Brazilian wandering spider.

Q: What arachnid can "web-swing" like a pendulum to escape predators by releasing silk threads while jumping? A: The jumping spider.

Q: Which spider species has an advanced reproductive strategy where the male deposits sperm into a web sac before mating? A: The tarantula.

Q: What arachnid produces venom with antimicrobial properties to prevent its prey from rotting before consumption? A: The yellow sac spider.

Q: Which species of spider uses its legs to create "wave-like" movements in its web to confuse prey? A: The giant house spider.

Q: What spider species builds "traplines" of silk to detect vibrations from prey approaching? A: The redback spider.

Q: Which arachnid species exhibits "maternal care," where the female protects her eggs and young? A: The wolf spider.

Q: What arachnid uses its spinnerets to create a specialized net to fling at prey? A: The ogre-faced spider.

Q: Which arachnid has evolved to detect the chemical signature of its prey using chemosensory hairs? A: The scorpion.

Q: What arachnid can control the viscosity of its silk, making it sticky or non-sticky depending on its function? A: The orb-weaver spider.

Q: Which spider can produce silk that absorbs water from the atmosphere to maintain its elasticity? A: The Darwin's bark spider.

Q: What arachnid is capable of creating intricate 3D webs to ensnare prey from multiple angles? A: The cobweb spider.

Q: Which arachnid has specialized eyes for detecting polarized light, allowing it to navigate even in low light conditions? A: The jumping spider.

Q: What spider uses "vibrational" signals to mimic the struggles of an insect trapped in another spider's web to lure the web's owner? A: The pirate spider.

Q: Which species of arachnid has evolved to hunt prey in total darkness using infrared detection? A: The pseudoscorpion.

Q: What species of spider has evolved to live in symbiosis with plants, protecting them from insect pests in exchange for shelter? A: The acacia ant spider.

Q: Which arachnid can survive extreme temperatures, ranging from sub-freezing to over 100°F (38°C)? A: The desert scorpion.

Q: What spider is known for laying "decoy" webs to trap and disorient its prey before delivering a fatal bite? A: The trapdoor spider.

Q: Which species of arachnid produces a specialized venom that acts as an anticoagulant in its prey? A: The brown recluse spider.

Q: What species of spider can sense the electrical fields produced by other insects to detect and capture them? A: The orb-weaver spider.

Now, let's look at the scientific names of several common arachnids.

Latrodectus mactans – Black Widow Spider
Loxosceles reclusa – Brown Recluse Spider
Parasteatoda tepidariorum – Common House Spider
Pholcus phalangioides – Cellar Spider (Daddy Longlegs)
Tegenaria domestica – Domestic House Spider
Argiope aurantia – Yellow Garden Spider
Eratigena atrica – Giant House Spider
Misumena vatia – Goldenrod Crab Spider
Araneus diadematus – European Garden Spider (Cross Orb-Weaver)
Aphonopelma chalcodes – Arizona Blonde Tarantula
Brachypelma smithi – Mexican Redknee Tarantula
Pardosa milvina – Thin-Legged Wolf Spider
Heterometrus laoticus – Asian Forest Scorpion
Centruroides sculpturatus – Arizona Bark Scorpion
Euscorpius italicus – Italian Scorpion
Galeodes arabs – Camel Spider (Solifuge)
Amblyomma americanum – Lone Star Tick

Ixodes scapularis – Deer Tick (Black-legged Tick)

Cheiracanthium inclusum – Yellow Sac Spider

Lycosidae – Wolf Spider (Family includes many common species)

Birds

Birds are a type of animal known for their feathers, which set them apart from all other creatures. Most birds are capable of flying, thanks to their wings and lightweight, hollow bones, although some, like ostriches and penguins, don't fly but are still classified as birds.

Another key feature of birds is their ability to lay eggs with hard shells. Birds also have beaks instead of teeth, and they use them for eating, building nests, or defending themselves. Their diets vary widely, with some birds eating insects, others eating seeds, fruits, or even small animals.

Birds are warm-blooded, which means they can maintain a constant body temperature, no matter the weather. They also have specialized respiratory systems that help them take in more oxygen, which is especially important when flying.

Birds are found in almost every environment on Earth, from forests and deserts to cities and oceans. They are also known for their songs and calls, which they use to communicate with each other. Some birds, like parrots and ravens, are highly intelligent, capable of problem-solving and even mimicking sounds.

Overall, birds are fascinating animals known for their feathers, wings, and beaks, and they play important roles in nature by pollinating plants, controlling insect populations, and spreading seeds.

Q: Which bird has the longest wingspan of any living bird? A: The wandering albatross, with a wingspan of up to 12 feet.

Q: What is the fastest bird in the world when in a dive? A: The peregrine falcon, reaching speeds of over 200 mph in a dive.

Q: Which bird is known for its elaborate courtship dance that includes "moonwalking"? A: The red-capped manakin.

Q: What is the only bird capable of hovering in place? A: The hummingbird.

Attracting hummingbirds to your yard is all about offering what they love—nectar, bright flowers, and a good place to perch. The easiest way to draw them in is with a hummingbird feeder filled with a simple homemade nectar (just mix four parts water to one part sugar). Skip the red dye—it's unnecessary and can be harmful.

Hummingbirds are also drawn to bright, tubular flowers, so plant species like bee balm, trumpet vine, and salvia. These flowers mimic the shape of their favorite blossoms, making your garden irresistible to them.

Adding some shady areas with trees or bushes will give them a place to rest between sips. Hummingbirds are surprisingly territorial, so having a few feeders spread out will help avoid fights over food.

Lastly, keeping your feeder clean and the nectar fresh is key. Dirty feeders can develop mold that harms these tiny birds. With the right setup, you'll have hummingbirds zipping around in no time!

Q: Which bird migrates the farthest distance annually? A: The Arctic tern, migrating up to 44,000 miles round trip.

Q: What bird species can sleep while flying? A: The common swift.

Q: Which bird is known for mimicking the sounds of other animals, including chainsaws and car alarms? A: The lyrebird.

Q: What bird has the ability to see ultraviolet light? A: The kestrel.

Q: Which bird builds its nest underwater? A: The dipper.

Q: What bird is known for having the largest number of feathers? A: The whistling swan, with over 25,000 feathers.

Q: Which bird is the only one with nostrils at the end of its beak? A: The kiwi.

Q: What is the heaviest flying bird in the world? A: The kori bustard.

Q: Which bird can rotate its head 270 degrees? A: The owl.

Q: What bird is known for being able to drink saltwater without getting sick? A: The albatross.

Q: Which bird has the largest beak in proportion to its body size? A: The toucan.

Q: What bird species mates for life and is often seen as a symbol of love? A: The swan.

Woodpeckers are a mixed bag—while they can be destructive to homes, they're incredibly beneficial to the ecosystem. When it comes to houses, woodpeckers can cause a lot of damage, especially if they mistake your siding or trim for a tree. They'll peck away, making holes in wood as they search for insects or create nesting sites. The sound alone can be annoying, but the damage to the exterior of homes can be costly to repair, especially if they keep coming back.

However, in the wild, woodpeckers are nature's pest control experts. They help keep insect populations in check by feeding on bugs burrowed deep into trees. Their pecking also creates nesting cavities that other birds and small animals use. Woodpeckers are important for forest health, and without them, insect infestations could cause bigger problems.

So, while they can be a pain for homeowners, they play a vital role in maintaining balanced ecosystems.

Q: Which bird is capable of running at speeds up to 43 mph? A: The ostrich.

Q: What bird can fly backward, forward, and sideways? A: The hummingbird.

Q: Which bird's heart can beat up to 1,200 times per minute? A: The hummingbird.

Q: What bird is known for having one of the most complex songs in the animal kingdom, lasting up to 10 minutes? A: The nightingale.

Q: Which bird species can remember human faces and hold grudges? A: The crow.

Q: What bird is known for its unique method of fishing by dropping bread or insects onto the water to lure fish? A: The green heron.

Q: Which bird migrates from one hemisphere to the other to avoid winter, traveling thousands of miles annually? A: The bar-tailed godwit.

Q: What bird has the largest brain relative to its body size? A: The New Caledonian crow.

Q: Which bird can fly nonstop for up to 10 months without landing? A: The common swift.

Q: What bird has the longest tongue relative to its body size? A: The woodpecker.

Q: Which bird is known for impaling its prey on thorns or barbed wire? A: The shrike.

Q: What bird can travel at night using the stars to navigate? A: The indigo bunting.

Q: Which bird lays the largest egg in proportion to its body size? A: The kiwi.

Q: What bird species has the best night vision? A: The owl.

Q: Which bird is known for its ability to "sing" with its wings? A: The club-winged manakin.

Q: What bird has a nesting strategy that involves laying its eggs in the nests of other bird species? A: The common cuckoo.

Q: Which bird species can mimic human speech and other sounds, sometimes even better than parrots? A: The African grey parrot.

Q: What bird is the smallest in the world? A: The bee hummingbird.

Q: Which bird is known for having blue feet and performing a unique mating dance? A: The blue-footed booby.

Q: What bird holds the record for the longest migratory flight without stopping? A: The bar-tailed godwit.

Q: Which bird's wings make a whistling sound during flight? A: The mourning dove.

Q: What bird species is capable of "sunbathing" to regulate its body temperature? A: The turkey vulture.

Q: Which bird uses its strong beak to break open bones and access marrow? A: The bearded vulture.

Q: What bird has the largest wingspan relative to its body weight? A: The frigatebird.

Q: Which bird is known for its cooperative breeding behavior, where offspring help raise their younger siblings? A: The acorn woodpecker.

Q: What bird has a call that can be heard over a mile away in dense forest? A: The howler monkey (wait, that's a mammal, but the bird that has a loud call is the screaming piha).

Q: Which bird species has no flight muscles and relies on gliding to get around? A: The kakapo.

Q: What bird is famous for building elaborate "bowers" to attract mates? A: The bowerbird.

Q: Which bird is known for traveling up to 60,000 miles per year, more than any other bird? A: The sooty shearwater.

Q: What bird has the thickest feathers of any species? A: The emperor penguin.

Q: Which bird can change the color of its feathers to blend in with its environment during different seasons? A: The willow ptarmigan.

Q: What bird uses its beak to "drum" on trees, sometimes over 10,000 times a day? A: The woodpecker.

Q: Which bird species can dive to depths of over 600 feet? A: The emperor penguin.

Q: What bird is known for performing "sky-dances" during courtship? A: The skylark.

Q: Which bird species uses tools to extract insects from tree bark? A: The New Caledonian crow.

Q: What bird can produce two different songs simultaneously? A: The northern mockingbird.

Q: Which bird builds the largest nests of any bird species? A: The bald eagle.

Q: What bird's heart can stop beating temporarily when it dives deep underwater? A: The common murre.

Q: Which bird can fly up to 56 mph in level flight? A: The red-breasted merganser.

Q: What bird has the fastest wingbeat of any species, reaching up to 80 beats per second? A: The ruby-throated hummingbird.

Q: Which bird species has been found to migrate during the day using the sun for navigation? A: The European starling.

Q: What bird can hear prey moving underground, allowing it to hunt by sound alone? A: The barn owl.

Q: Which bird lays the smallest egg relative to its size? A: The ostrich.

Q: What bird species is known for its unusual "umbrella" feathers that it flares during courtship displays? A: The umbrella bird.

Q: Which bird is famous for forming massive flocks known as murmurations? A: The European starling.

Q: What bird is capable of seeing magnetic fields to help it navigate during migration? A: The European robin.

Q: Which bird can swallow its food whole, including fish larger than its head? A: The pelican.

Q: What bird has a "tongue" that extends out of its skull and wraps around the back of its head? A: The woodpecker.

Q: Which bird has the longest lifespan, living up to 70 years in the wild? A: The albatross.

Q: What bird species is known for regurgitating food to feed its chicks? A: The penguin.

Q: Which bird is capable of flying in the thin air of the Himalayan mountains? A: The bar-headed goose.

Q: What bird has the ability to mimic the calls of over 20 different bird species? A: The northern mockingbird.

Q: Which bird can store food in its throat pouch for later consumption? A: The pelican.

Q: What bird is capable of sleeping with one eye open, keeping half its brain awake to watch for predators? A: The mallard duck.

Q: Which bird has been recorded traveling at 67 mph in level flight? A: The grey-headed albatross.

Q: What bird can dive from great heights, striking the water at over 80 mph to catch prey? A: The gannet.

Q: Which bird is known for having highly developed spatial memory to remember the locations of hidden food caches? A: The Clark's nutcracker.

Q: What bird can mimic the alarm calls of other species to scare them away from food? A: The drongo.

Q: Which bird lays the largest egg of any bird species relative to body size? A: The kiwi.

Q: What bird is known for its iridescent feathers, which appear to change color depending on the angle of light? A: The peacock.

Q: Which bird species is the largest member of the crow family?
A: The common raven.

Q: What bird can dive the deepest of any bird, reaching depths of over 1,800 feet? A: The emperor penguin.

Q: Which bird is known for "wing clapping" as part of its courtship display? A: The ruffed grouse.

Q: What bird can fly upside down for short distances? A: The hummingbird.

Q: Which bird species has a distinctive "U" shaped flight pattern during migration? A: The barn swallow.

Q: What bird's feathers contain melanin, making them more resistant to wear and tear? A: The crow.

Q: Which bird is known to use spider silk to weave its nest? A: The hummingbird.

Q: What bird species uses its sharp beak to break through the hard shells of snails? A: The song thrush.

Q: Which bird can communicate using non-vocal sounds, such as wing beats or tail feathers? A: The ruffed grouse.

Q: What bird has a "waterproof" coating on its feathers to help it stay dry in water? A: The penguin.

Q: Which bird species can carry stones in its beak to weigh itself down while walking underwater? A: The dipper.

Q: What bird has the most complex vocal range of any species? A: The lyrebird.

Q: Which bird uses "anting," a behavior where it rubs ants on its feathers to deter parasites? A: The jay.

Q: What bird species performs elaborate aerial displays, including somersaults, to attract a mate? A: The skylark.

Q: Which bird's call has been described as sounding like a human laugh? A: The kookaburra.

Q: What bird species can "see" Earth's magnetic field to help it migrate? A: The European robin.

Q: Which bird is known for its distinctive "laughing" call? A: The laughing gull.

Q: What bird can drink nectar while hovering? A: The hummingbird.

Q: Which bird is known for creating "rafts" of floating vegetation to support its nests? A: The African jacana.

Q: What bird can fly at altitudes up to 37,000 feet? A: The bar-headed goose.

Q: Which bird can hunt fish by diving headfirst into the water from great heights? A: The osprey.

Q: What bird uses its large bill to catch prey and toss it into the air before swallowing? A: The toucan.

Q: Which bird species mates for life and is known for their strong pair bonds? A: The bald eagle.

Q: What bird's diet is primarily composed of bones, which it digests using powerful stomach acid? A: The bearded vulture.

Advanced Fun Facts

Q: Which bird is capable of "tapping" into tree sap and returning to drink it later? A: The yellow-bellied sapsucker.

Q: What bird species is known for its ability to fly over the Himalayas during migration, reaching altitudes of up to 29,000 feet? A: The bar-headed goose.

Q: Which bird uses "tools" by stripping leaves from twigs to probe for insects? A: The New Caledonian crow.

Q: What bird species has the longest migratory flight made by a land bird, traveling over 8,000 miles? A: The pectoral sandpiper.

Q: Which bird's eggs are known for being camouflaged with speckled patterns to blend into the environment? A: The killdeer.

Q: What bird species uses its bill to drum out messages on hollow objects, similar to Morse code? A: The woodpecker.

Q: Which bird can see in the ultraviolet spectrum, allowing it to detect rodent urine trails? A: The kestrel.

Q: What bird has a crop capable of storing food and later regurgitating it to feed chicks? A: The pigeon.

Q: Which bird is known for "displacing" other birds by laying its eggs in their nests, leaving them to raise its chicks? A: The common cuckoo.

Q: What bird species has a specialized beak adapted for feeding on pine seeds extracted from cones? A: The crossbill.

Q: Which bird can fly up to 13,000 feet above sea level and navigate over long distances using Earth's magnetic fields? A: The homing pigeon.

Q: What bird can see at speeds faster than humans can blink, detecting movements of prey in flight? A: The peregrine falcon.

Q: Which bird's feathers are structurally colored rather than pigmented, giving them iridescent properties? A: The peacock.

Q: What bird is known for being an "obligate" brood parasite, meaning it never builds its own nest? A: The brown-headed cowbird.

Q: Which bird species exhibits "altruistic behavior" by helping unrelated individuals raise their offspring? A: The Florida scrub jay.

Q: What bird uses echolocation in dark caves to navigate, a rare ability among birds? A: The oilbird.

Q: Which bird is capable of using its sharp talons to catch fish in mid-air as they leap out of the water? A: The osprey.

Q: What bird species has been observed "bait fishing" by using objects like bread or leaves to lure fish closer? A: The green heron.

Q: Which bird has one of the largest flight ranges for a non-migratory species, covering up to 50 miles in a single day? A: The wandering albatross.

Q: What bird has the highest number of known vocalizations, with up to 1,000 different sounds? A: The northern mockingbird.

Q: Which bird species uses thermals, or columns of rising air, to soar for hours without flapping its wings? A: The turkey vulture.

Q: What bird is capable of detecting air pressure changes, allowing it to predict incoming storms? A: The golden plover.

Q: Which bird is known for using "leaf umbrellas" to shade itself from the sun during foraging? A: The black heron.

Q: What bird has the smallest heart relative to its body size among flying birds? A: The ostrich.

Q: Which bird species can fly backward, forward, and hover while consuming nectar from flowers? A: The hummingbird.

Q: What bird is capable of mimicking human speech as well as the sounds of other birds, animals, and mechanical noises? A: The African grey parrot.

Q: Which bird is known for creating "raft" formations on the water's surface during migration to conserve energy? A: The common eider.

Q: What bird is famous for its cooperative hunting strategy, where flocks work together to herd fish or other prey? A: The Harris's hawk.

Q: Which bird's feathers contain melanin, making them more resistant to wear and tear from sunlight exposure? A: The raven.

Q: What bird can sleep while perched on one leg due to a locking mechanism in its tendons? A: The flamingo.

Q: Which bird species has a specialized oil gland used to waterproof its feathers during grooming? A: The penguin.

Q: What bird can hover in one spot for extended periods while searching for prey, a behavior known as "kiting"? A: The kestrel.

Q: Which bird uses its long tail feathers to stabilize itself while climbing vertically on tree trunks? A: The woodpecker.

Q: What bird species uses "solar navigation," orienting itself with the position of the sun during migration? A: The homing pigeon.

Q: Which bird species engages in "sky-pointing" as a courtship display? A: The frigatebird.

Q: What bird has a tongue that can extend up to three times the length of its bill? A: The northern flicker.

Q: Which bird species builds its nest entirely out of saliva, which hardens into a solid structure? A: The edible-nest swiftlet.

Q: What bird species is known for creating "fishing holes" in ice-covered ponds to catch fish? A: The bald eagle.

Q: Which bird species has been known to engage in cooperative nest building, where multiple individuals contribute to a communal nest? A: The sociable weaver.

Q: What bird uses its sharp beak to peck holes in the bark of trees, creating "sap wells" for feeding? A: The red-naped sapsucker.

Q: Which bird species forms life-long pair bonds and performs synchronized swimming as part of its courtship ritual? A: The black swan.

Q: What bird has a specialized syrinx, or vocal organ, allowing it to produce multiple notes simultaneously? A: The song thrush.

Q: Which bird species has the most efficient respiratory system of any vertebrate, allowing for continuous airflow during flight? A: The pigeon.

Q: What bird uses its tail as a rudder while flying, providing stability during tight maneuvers? A: The barn swallow.

Q: Which bird species nests in vertical sandbanks and uses its sharp claws to dig burrows? A: The sand martin (bank swallow).

Q: What bird species can memorize and reproduce the songs of over 50 other bird species? A: The northern mockingbird.

Q: Which bird has a specialized sense of smell, allowing it to detect food sources from miles away? A: The turkey vulture.

Q: What bird is known for having one of the largest nests of any bird species, with nests weighing up to a ton? A: The bald eagle.

Q: Which bird can alter its flight path by shifting the angle of its feathers mid-air to catch thermals? A: The condor.

Q: What bird uses its long, curved beak to extract insects from tree bark, specializing in deep probing? A: The treecreeper.

Okay. Let's get geeky with a few scientific names for common birds.

Pelecanus occidentalis – Brown Pelican
Aquila chrysaetos – Golden Eagle
Falco peregrinus – Peregrine Falcon
Columba livia – Rock Dove (Pigeon)
Sturnus vulgaris – European Starling
Turdus migratorius – American Robin
Cardinalis cardinalis – Northern Cardinal
Struthio camelus – Common Ostrich

Pica pica – Eurasian Magpie
Corvus corax – Common Raven
Apus apus – Common Swift
Cygnus olor – Mute Swan
Bubo virginianus – Great Horned Owl
Haliaeetus leucocephalus – Bald Eagle
Phoenicopterus roseus – Greater Flamingo
Pica hudsonia – Black-billed Magpie
Passer domesticus – House Sparrow
Anas platyrhynchos – Mallard Duck
Troglodytes aedon – House Wren
Cyanocitta cristata – Blue Jay

Reptiles

Reptiles are cold-blooded animals, which means they rely on the temperature of their surroundings to stay warm. Unlike mammals and birds, reptiles cannot regulate their own body temperature, so they need to bask in the sun to warm up or find shade to cool down.

Reptiles have scaly skin that helps protect them and prevent water loss, which is why many reptiles are well-suited to dry environments. Most reptiles lay leathery-shelled eggs, though some, like certain species of snakes, give birth to live young.

One of the key traits of reptiles is that they have lungs and need to breathe air. They come in a variety of forms and sizes, but most reptiles have four legs or none at all, like snakes. There are four main groups of reptiles:

1. Lizards and snakes (like geckos and pythons),
2. Turtles and tortoises (like sea turtles and land tortoises),
3. Crocodilians (like crocodiles and alligators), and
4. Tuataras, a rare type of reptile found in New Zealand.

Reptiles are carnivorous or omnivorous, meaning they eat meat, plants, or both. Some, like snakes, use venom to hunt their prey, while others, like turtles, rely on strong jaws or claws.

Reptiles are found in a variety of habitats, from deserts and forests to swamps and oceans. They are ancient creatures, having existed since the time of the dinosaurs.

Q: Which reptile is known for being the fastest on land? A: The black spiny-tailed iguana, reaching speeds up to 21 mph.

Q: What is the largest species of crocodile? A: The saltwater crocodile.

Q: Which reptile has a third "parietal eye" on the top of its head? A: The tuatara.

Q: What reptile can change its skin color to regulate body temperature or communicate with others? A: The chameleon.

Q: Which species of turtle can live for over 150 years? A: The Galápagos tortoise.

Q: What is the longest venomous snake in the world? A: The king cobra.

Q: Which reptile is known to be able to run on water? A: The basilisk lizard (often called the "Jesus lizard").

Q: What reptile can detach and regrow its tail as a defense mechanism? A: The gecko.

Q: Which species of snake can "fly" by flattening its body and gliding between trees? A: The paradise tree snake.

Q: What is the heaviest species of snake? A: The green anaconda.

Q: Which reptile species is known for playing dead when threatened? A: The hognose snake.

Q: What reptile species has no external ear openings and relies on vibrations to detect sound? A: The snake.

Q: Which reptile can breathe through its cloaca when underwater for long periods? A: The Fitzroy River turtle.

Q: What is the only lizard species capable of true vocalization? A: The tokay gecko.

Q: Which reptile can survive for months without food due to its slow metabolism? A: The Komodo dragon.

Q: What reptile can inflate its body to make it appear larger to predators? A: The chuckwalla.

Q: Which snake has the longest fangs of any venomous species? A: The gaboon viper.

Q: What reptile species is capable of using chemical communication through pheromones? A: The skink.

Q: Which reptile is known for its ability to drink water through its skin? A: The thorny devil.

Q: What is the only venomous lizard species in North America?
A: The Gila monster.

Q: Which species of reptile can live its entire life without drinking liquid water, obtaining moisture from food? A: The desert tortoise.

Q: What is the smallest species of snake in the world? A: The Barbados threadsnake.

Q: Which reptile can detect infrared radiation, allowing it to sense the body heat of prey? A: The pit viper.

Pit vipers, especially rattlesnakes, are known for their distinctive rattle sound—a pretty clear warning that they're nearby and not to be messed with! But a common myth is that you can tell a rattlesnake's age by counting the segments of its rattle. In reality, each time a rattlesnake sheds its skin (which can happen a few times a year), it adds a new segment to the rattle. So, while older snakes tend to have longer rattles, you can't really use it as a perfect age indicator.

Rattles can also break off, so a shorter rattle doesn't necessarily mean a young snake. The rattle itself is made of keratin (the same stuff as your fingernails), and the sound is produced when the segments knock together.

As pit vipers, rattlesnakes have heat-sensing pits on their heads that help them detect warm-blooded prey. They're fascinating creatures—just keep your distance and appreciate that rattle as the polite warning it's meant to be!

Q: What reptile species has been observed using tools, such as rocks, to crack open eggs? A: The Komodo dragon.

Q: Which species of snake is the most aggressive in the world? A: The black mamba.

Q: What reptile can shoot blood from its eyes as a defense mechanism? A: The horned lizard.

Q: Which reptile has teeth that can slice through bone with ease? A: The Nile crocodile.

Q: What reptile uses its long, forked tongue to "taste" the air and detect chemical signals? A: The snake.

Q: Which species of lizard can walk on sand dunes without sinking due to fringed toes? A: The fringe-toed lizard.

Q: What reptile has one of the most powerful immune systems, allowing it to heal quickly from infections? A: The Komodo dragon.

Q: Which reptile can change its sex from male to female depending on temperature? A: The bearded dragon.

Q: What reptile has specialized scales that allow it to store water for long periods? A: The desert tortoise.

The desert tortoise is a fascinating creature, known for its slow pace and impressively long lifespan. These tortoises, native to the deserts of the southwestern United States and northern

Mexico, can live for 50 to 80 years in the wild, sometimes even longer in captivity. They've evolved to thrive in harsh desert environments, spending much of their time in burrows to escape the extreme heat.

One of the oldest known desert tortoises was a tortoise named Gertie, who lived to be 100 years old! Gertie lived in captivity, where she enjoyed a more protected life compared to her wild counterparts.

Desert tortoises grow slowly, maturing over several decades, and their long lifespan is part of what makes them so special. Sadly, they're considered a vulnerable species due to habitat loss, so efforts are being made to protect these amazing animals and ensure they keep roaming the deserts for many more years!

Q: Which reptile can go into a state of hibernation in cold temperatures, slowing its heartbeat to survive? A: The painted turtle.

Q: What is the most venomous snake in the world? A: The inland taipan.

Q: Which reptile has a sticky tongue that can extend to catch insects? A: The chameleon.

Q: What reptile can produce venom that causes severe bleeding in its prey? A: The boomslang snake.

Q: Which species of reptile is known for its ability to climb vertical walls and even glass? A: The gecko.

Q: What reptile has the ability to survive for long periods underwater without surfacing for air? A: The American alligator.

Q: Which reptile is known for "spitting" venom to blind or deter its enemies? A: The spitting cobra.

Q: What species of lizard has frilled skin around its neck that it expands to appear larger when threatened? A: The frilled lizard.

Q: Which reptile can dig complex underground burrows to escape extreme temperatures? A: The gopher tortoise.

Q: What is the longest species of lizard? A: The Komodo dragon.

Q: Which reptile can hold its breath underwater for over two hours? A: The marine iguana.

Q: What species of snake is known for its ability to climb trees and move across branches? A: The green tree python.

Q: Which reptile has a forked tongue that functions similarly to a snake's for detecting prey? A: The monitor lizard.

Q: What species of reptile can live for over 50 years in captivity? A: The ball python.

Q: Which reptile has a bite force stronger than that of a great white shark? A: The saltwater crocodile.

Q: What is the only species of reptile that can fly or glide through the air? A: The Draco lizard (flying dragon).

Q: Which species of snake can swim and hunt both in freshwater and saltwater environments? A: The sea snake.

Q: What reptile is the fastest swimmer, reaching speeds of up to 15 mph in water? A: The leatherback sea turtle.

Q: Which reptile has been found to use vocal communication, including grunts and hisses, to interact with others? A: The American alligator.

Q: What reptile can consume prey up to 80% of its body weight in a single meal? A: The Burmese python.

Q: Which species of lizard uses "head-bobbing" as a territorial display? A: The bearded dragon.

Q: What reptile is known for having one of the longest lifespans, often living over 100 years in the wild? A: The Aldabra giant tortoise.

Q: Which species of reptile is entirely herbivorous despite being closely related to carnivorous species? A: The green iguana.

Q: What species of lizard can lose and regrow its tail up to five times in its lifetime? A: The anole lizard.

Q: Which reptile has specialized heat-sensing pits on its face to locate warm-blooded prey? A: The rattlesnake.

Q: What reptile has transparent lower eyelids that allow it to see underwater? A: The crocodile.

Q: Which species of snake is known for being able to coil and strike with extreme speed? A: The black mamba.

Q: What reptile can survive being frozen solid during hibernation and "thaw" out in the spring? A: The wood frog (though technically an amphibian, its behavior is similar to reptiles).

Q: Which reptile is known for having some of the longest venom fangs relative to body size? A: The gaboon viper.

Q: What reptile has a specialized jaw hinge that allows it to open its mouth wider than its head? A: The snake.

Q: Which species of lizard is known for its ability to "run" across hot sand without burning its feet? A: The zebra-tailed lizard.

Q: What reptile is known to sunbathe to regulate its body temperature? A: The crocodile.

Q: Which reptile is known for its ability to remain perfectly still for hours while hunting? A: The Komodo dragon.

Q: What species of snake is known to be able to survive months without food due to its slow metabolism? A: The ball python.

Q: Which reptile has transparent scales covering its eyes, as it has no eyelids? A: The snake.

Q: What species of turtle can "breathe" through its skin, especially in cold water? A: The painted turtle.

Q: Which species of reptile is known for forming "breeding balls," where multiple males surround a single female? A: The garter snake.

Q: What species of reptile is known to use ultraviolet light for communication? A: The anole lizard.

Q: Which species of snake gives birth to live young instead of laying eggs? A: The boa constrictor.

Q: What species of reptile can go into a state of "brumation," a form of hibernation in cold temperatures? A: The bearded dragon.

Q: Which reptile species has been found to use problem-solving abilities to escape enclosures? A: The monitor lizard.

Q: What species of snake has the ability to flatten its neck to appear larger and more intimidating? A: The cobra.

Q: Which species of reptile can change the color of its skin to communicate mood or attract mates? A: The chameleon.

Q: What reptile is known for digging deep burrows to escape the desert heat? A: The desert tortoise.

Q: Which species of snake can hold its breath underwater for up to 30 minutes while hunting fish? A: The cottonmouth (water moccasin).

Q: What reptile can climb trees using sharp claws and a prehensile tail for balance? A: The green iguana.

Q: Which species of reptile is known for its ability to regrow entire limbs, including toes? A: The axolotl (though technically an amphibian, often confused with reptiles).

Q: What species of reptile has specialized "gliding" skin folds that allow it to soar from tree to tree? A: The flying dragon (Draco lizard).

Q: Which species of reptile is capable of producing venom from glands located in its lower jaw? A: The beaded lizard.

Q: What reptile uses its spiny tail as a weapon, swinging it to fend off predators? A: The uromastyx lizard.

Q: Which species of lizard is known for inflating its body to prevent predators from pulling it out of crevices? A: The chuckwalla.

Q: What reptile has specialized claws for digging complex burrows to protect eggs? A: The snapping turtle.

Q: Which reptile species can grow to over 20 feet in length and weigh more than 1,000 pounds? A: The saltwater crocodile.

Q: What reptile can "puff up" its body and hiss to appear larger and more threatening? A: The Gila monster.

The Gila monster is one of the few venomous lizards in the world, and while it's not aggressive, its bite can be toxic to humans. Found in the southwestern United States and northern Mexico, the Gila monster's venom isn't injected like a snake bite; instead, it flows from glands in its lower jaw and seeps into the wound as the lizard chews. The bite itself can be painful, and the venom can cause symptoms like intense burning pain, swelling, and nausea.

Luckily, Gila monster bites are rarely life-threatening to humans, but they're definitely not something you want to experience! These lizards use their venom mainly for defense and to subdue small prey, not to attack people. The good news is, they're slow-moving and tend to avoid humans, so bites are pretty rare. If you ever encounter a Gila monster, it's best to admire it from a safe distance—no need to get up close and personal!

Q: Which species of reptile has a lifespan that can exceed 100 years in captivity? A: The Aldabra giant tortoise.

Q: What reptile is known to form "hunting parties" to catch prey, a rare behavior in reptiles? A: The Komodo dragon.

Q: Which species of lizard is known for its ability to "run" on two legs when threatened? A: The basilisk lizard.

Q: What reptile has specialized "conical" teeth for gripping and tearing prey? A: The crocodile.

Q: Which species of reptile uses its long, sticky tongue to capture insects at high speed? A: The chameleon.

Q: What reptile can survive for weeks without food by storing fat in its tail? A: The leopard gecko.

Q: Which species of snake is known to "strike" with incredible speed, delivering a venomous bite in milliseconds? A: The pit viper.

Q: What reptile is capable of changing its body shape to "fit" into tight spaces when hiding from predators? A: The glass lizard.

Q: Which species of turtle has flippers adapted for swimming long distances in the ocean? A: The leatherback sea turtle.

Q: What species of lizard is known for its ability to drop its tail as a distraction and then grow a new one? A: The skink.

Q: Which species of snake is capable of mimicking the sound of a rattlesnake by vibrating its tail? A: The kingsnake.

Q: What reptile is known for its ability to strike prey with precision, using heat-sensing pits on its face? A: The rattlesnake.

Q: Which species of lizard uses its sharp claws and strong jaws to tear apart its prey? A: The Komodo dragon.

Q: What species of reptile has been found to engage in "parental care," guarding its eggs and young after hatching? A: The American alligator.

Advanced Fun Facts

Q: Which reptile species is known for producing both venom and an anticoagulant in its bite, causing excessive bleeding in its prey? A: The boomslang (Dispholidus typus).

Q: What reptile has the highest bite force of any living animal, capable of generating up to 3,700 PSI? A: The saltwater crocodile (Crocodylus porosus).

Q: Which species of snake is unique in that it gives birth to live young rather than laying eggs? A: The boa constrictor (Boa constrictor).

Q: What reptile species is known for being able to detect ultraviolet light, which aids in foraging and communication? A: The anole lizard (Anolis spp.).

Q: Which species of turtle can survive the freezing of up to 50% of its body water during winter hibernation? A: The painted turtle (Chrysemys picta).

Q: What is the only known venomous turtle species? A: The softshell turtle (Pelodiscus sinensis) has venom glands, but their toxicity is minimal to humans.

Q: Which species of lizard exhibits temperature-dependent sex determination, where the incubation temperature of eggs determines the sex of the hatchlings? A: The American alligator (Alligator mississippiensis).

Q: What reptile is known for its ability to drink through its skin by capturing water from rain or dew? A: The thorny devil (Moloch horridus).

Q: Which species of snake can glide through the air by flattening its body and undulating as it "flies" from tree to tree? A: The paradise tree snake (Chrysopelea paradisi).

Q: What species of reptile has the ability to deliver a potentially lethal bite even after being decapitated, due to reflexive muscle contractions? A: The rattlesnake (Crotalus spp.).

Q: Which species of lizard can communicate using color changes to signal territorial disputes or mating intentions? A: The panther chameleon (Furcifer pardalis).

Q: What is the only known species of lizard that secretes venom from glands located in its lower jaw? A: The Gila monster (Heloderma suspectum).

Q: Which species of turtle has been known to "hibernate" underwater for extended periods during cold seasons, absorbing oxygen through its cloaca? A: The Fitzroy River turtle (Rheodytes leukops).

Q: What species of snake has evolved specialized scales on its belly to help it climb trees vertically? A: The green tree python (Morelia viridis).

Q: Which species of crocodile has developed an adaptation that allows it to survive for months without eating during the dry season? A: The Nile crocodile (Crocodylus niloticus).

Q: What reptile is known for its ability to "triploid" reproduction, where females produce offspring without mating, a form of parthenogenesis? A: The New Mexico whiptail lizard (Aspidoscelis neomexicanus).

Q: Which species of snake is known to coil its prey in constriction with such force that it can stop the prey's heart almost immediately? A: The green anaconda (Eunectes murinus).

Q: What species of turtle has the largest migration range, traveling thousands of miles between feeding and nesting grounds? A: The leatherback sea turtle (Dermochelys coriacea).

Q: Which reptile species is known for its "egg-eating" behavior, where it uses specialized vertebrae in its throat to crack the eggshells? A: The egg-eating snake (Dasypeltis spp.).

Q: What species of snake is known for its ability to "play dead" by rolling onto its back and releasing a foul odor from its cloaca? A: The hognose snake (Heterodon spp.).

Q: Which reptile species uses its tail as a weapon, capable of delivering a forceful blow to predators or rivals? A: The monitor lizard (Varanus spp.).

Q: What species of snake has the ability to detect minute changes in barometric pressure, which helps it anticipate and avoid floods or storms? A: The coastal taipan (Oxyuranus scutellatus).

Q: Which species of reptile uses a specialized "Jacobson's organ" to enhance its sense of smell, especially during hunting or mating? A: The Komodo dragon (Varanus komodoensis).

Q: What reptile has a highly specialized hunting strategy, where it stalks prey in water for hours and strikes with lightning speed when within range? A: The saltwater crocodile (Crocodylus porosus).

Q: Which species of snake has venom that specifically targets the nervous system, resulting in rapid paralysis of its prey? A: The inland taipan (Oxyuranus microlepidotus).

Q: What species of reptile can walk on water thanks to its large, fringed toes that distribute its weight over the water's surface? A: The common basilisk (Basiliscus basiliscus).

Q: Which reptile has the ability to "brumate," entering a state of dormancy similar to hibernation during the winter months? A: The bearded dragon (Pogona vitticeps).

Q: What species of turtle has been found to use the Earth's magnetic fields to navigate during its long migrations? A: The loggerhead sea turtle (Caretta caretta).

Q: Which reptile species has evolved to live in urban environments and thrive on a diet of insects and small mammals around human settlements? A: The tokay gecko (Gekko gecko).

Q: What species of snake uses its heat-sensing pits to track down warm-blooded prey in complete darkness? A: The pit viper (Crotalinae subfamily).

Q: Which species of lizard can "see" polarized light, allowing it to navigate and find food in bright, reflective environments like deserts? A: The fringe-toed lizard (Uma spp.).

Q: What species of reptile has been observed using teamwork to catch larger prey, one of the few examples of cooperative hunting in reptiles? A: The Komodo dragon (Varanus komodoensis).

Q: Which species of snake has the longest recorded venomous fangs, measuring over 2 inches in length? A: The gaboon viper (Bitis gabonica).

Q: What species of reptile has specialized adaptations to drink "fog" water by collecting it on its scales and channeling it to its mouth? A: The Namib desert beetle (though technically not a reptile, some desert reptiles exhibit similar water-gathering behavior).

Q: Which reptile species has a flap of skin under its throat called a dewlap, which is used for communication and display? A: The green anole (Anolis carolinensis).

Q: What species of lizard can drop its tail when threatened, a process called "caudal autotomy," to escape predators? A: The skink (Scincidae family).

Q: Which reptile species has a highly specialized diet consisting mostly of ants and termites? A: The thorny devil (Moloch horridus).

Q: What species of snake has venom that contains hemotoxins, which destroy blood cells and cause internal bleeding? A: The boomslang (Dispholidus typus).

Q: Which species of lizard has "spectacles" covering its eyes, which act as a protective layer since it has no eyelids? A: The gecko (Gekkonidae family).

Q: What species of reptile has been found to communicate using complex vocalizations, including grunts, hisses, and roars? A: The American alligator (Alligator mississippiensis).

Q: Which species of turtle can retract both its head and limbs entirely inside its shell for protection? A: The box turtle (Terrapene spp.).

Q: What species of lizard is known for its "push-up" displays, which it uses to assert dominance and attract mates? A: The desert iguana (Dipsosaurus dorsalis).

Q: Which species of snake has a diet primarily consisting of other snakes, including venomous species? A: The king snake (Lampropeltis spp.).

Q: What species of crocodile has been known to "balance" sticks on its snout to lure nesting birds looking for twigs? A: The mugger crocodile (Crocodylus palustris).

Q: Which reptile is capable of walking considerable distances over land between water sources, especially in drought conditions? A: The saltwater crocodile (Crocodylus porosus).

Q: What species of snake is known for its "sidewinding" locomotion, allowing it to move efficiently across loose desert sand? A: The sidewinder rattlesnake (Crotalus cerastes).

Q: Which reptile species can lay eggs that remain dormant for months, only hatching when conditions become favorable? A: The fence lizard (Sceloporus spp.).

Q: What species of reptile uses "tail flicking" as a warning signal to potential predators? A: The rattlesnake (Crotalus spp.).

Q: Which reptile has the unique ability to regenerate not only its tail but also parts of its spinal cord? A: The axolotl (though technically an amphibian, it shares regenerative traits with some reptiles).

Q: What species of lizard is known for its ability to balance on two legs while running at high speeds to escape predators? A: The common basilisk (Basiliscus basiliscus).

Wait, there's more. Here are some scientific names for common reptiles yu may recognize.

Crocodylus porosus – Saltwater Crocodile
Alligator mississippiensis – American Alligator
Chelonia mydas – Green Sea Turtle
Testudo graeca – Greek Tortoise
Varanus komodoensis – Komodo Dragon
Pogona vitticeps – Central Bearded Dragon
Pantherophis guttatus – Corn Snake
Python regius – Ball Python
Eunectes murinus – Green Anaconda

Hemidactylus frenatus – Common House Gecko
Iguana iguana – Green Iguana
Chamaeleo calyptratus – Veiled Chameleon
Naja naja – Indian Cobra
Crotalus atrox – Western Diamondback Rattlesnake
Sceloporus occidentalis – Western Fence Lizard
Lacerta agilis – Sand Lizard
Sphenodon punctatus – Tuatara
Gekko gecko – Tokay Gecko
Dipsosaurus dorsalis – Desert Iguana
Tiliqua scincoides – Blue-tongued Skink

Amphibians

Amphibians are animals that live part of their life in water and part on land. They are cold-blooded, which means their body temperature changes with their environment. Most amphibians start life as aquatic larvae (like tadpoles), breathing through gills and living in water. As they grow, they go through a transformation called metamorphosis. For example, tadpoles grow into frogs and develop lungs to breathe air and legs for moving on land.

Amphibians have moist, smooth skin that helps them absorb water and oxygen, which is why they often live in wet or damp environments. Many amphibians can also breathe through their skin, though they still have lungs as adults.

There are three main types of amphibians:

1. Frogs and toads, known for their jumping legs and croaking sounds.
2. Salamanders and newts, which look like lizards but have moist skin and spend more time in water.

3. Caecilians, which are legless and look more like worms or snakes.

Amphibians are often carnivorous, feeding on insects, worms, and other small animals. They play an important role in ecosystems by controlling insect populations and serving as food for other animals. However, many amphibians are sensitive to environmental changes, which is why pollution and habitat destruction can easily harm them.

Q: Which amphibian can survive being frozen solid in winter and then thaw out in spring? A: The wood frog.

Q: What is the largest amphibian in the world? A: The Chinese giant salamander.

Q: Which amphibian can breathe through both its lungs and skin? A: The common frog.

Q: What amphibian's skin secretes toxins that are used by indigenous people to poison their blowgun darts? A: The golden poison dart frog.

Q: Which amphibian undergoes metamorphosis from a tadpole into an adult? A: The common toad.

Q: What amphibian is capable of regenerating lost limbs, including legs and parts of its heart? A: The axolotl.

Q: Which amphibian is known for having no lungs, relying entirely on its skin for respiration? A: The lungless salamander.

Q: What is the smallest amphibian in the world? A: The Paedophryne amauensis frog.

Q: Which amphibian makes a loud, high-pitched croak known as a "chorus" during the breeding season? A: The spring peeper.

Q: What amphibian uses its sticky tongue to capture prey like insects and small invertebrates? A: The green tree frog.

Q: Which amphibian has a defense mechanism that includes curling into a ball and excreting a sticky, toxic substance? A: The fire salamander.

Q: What amphibian has the ability to change its color to blend into its surroundings? A: The European tree frog.

One of the oddest and most unique amphibians out there has to be the axolotl. This little creature, native to lakes in Mexico, is like something out of a sci-fi movie. What makes the axolotl so strange is that it never fully grows up—it stays in its juvenile stage its whole life, a process known as neoteny. While most amphibians, like frogs, go through metamorphosis (changing from tadpoles to adults), the axolotl skips that step, keeping its feathery gills and aquatic lifestyle.

But that's not all. Axolotls have a superpower: regeneration. They can regrow entire limbs, parts of their spinal cord, and even bits of their heart and brain! Scientists have been studying them for years to understand how they do it.

Axolotls are also pretty cute, with their wide grins and feathery gills, making them popular pets. Sadly, they're critically endangered in the wild, but they're thriving in captivity. Truly a one-of-a-kind amphibian!

Q: Which amphibian is known for walking on land for short periods, even though it primarily lives in water? A: The African clawed frog.

Q: What amphibian can carry its eggs on its back until they hatch? A: The Surinam toad.

Q: Which amphibian can survive without food for up to 10 years by entering a state of dormancy? A: The tiger salamander.

Q: What amphibian's skin contains glands that secrete antifungal and antibacterial compounds? A: The American bullfrog.

Q: Which amphibian has external gills that allow it to breathe underwater throughout its life? A: The axolotl.

Q: What amphibian's powerful legs allow it to leap up to 10 times its body length? A: The red-eyed tree frog.

Q: Which amphibian lays its eggs in moist soil or under logs instead of in water? A: The eastern red-backed salamander.

Q: What amphibian is known for its ability to sing, producing a melodious "ribbit" sound? A: The Pacific tree frog.

Q: Which amphibian is nocturnal and spends most of its day hidden under rocks or in burrows? A: The spotted salamander.

Q: What amphibian species is fully aquatic and never undergoes complete metamorphosis? A: The axolotl.

Q: Which amphibian secretes a potent skin toxin that can kill predators, including birds and mammals? A: The cane toad.

Q: What amphibian hatches from its egg directly into a small frog, bypassing the tadpole stage entirely? A: The coqui frog.

Q: Which amphibian can project its tongue to catch prey in less than a second? A: The horned frog.

Q: What amphibian is capable of vocalizing underwater using air stored in its lungs? A: The African dwarf frog.

Q: Which amphibian has teeth in both its upper and lower jaws? A: The African clawed frog.

Q: What amphibian can secrete a slime that makes it slippery and difficult for predators to catch? A: The common newt.

The newt has long been associated with magic and witchcraft, famously appearing in witches' brews, like in Shakespeare's Macbeth, where the witches chant, "Eye of newt and toe of frog." But despite the spooky association, "eye of newt" wasn't literal—it was likely an old herbal term for mustard seed or another plant, as witches often used natural ingredients for their concoctions.

Newts, with their slimy skin and regenerative abilities (they can regrow lost limbs), might have seemed mystical to people in the past. Their amphibious nature—living both in water and on land—added to their mysterious aura, making them a fitting symbol for transformation, a key theme in magic.

Though modern witches and folklore enthusiasts now know better, the newt's connection to witchcraft persists in popular culture, lending a bit of creepy charm to the idea of bubbling cauldrons and magical brews. In reality, newts are just cool little creatures with fascinating biology!

Q: Which amphibian can lay up to 20,000 eggs at a time in water? A: The American bullfrog.

Q: What amphibian has adapted to living in the rainforest canopy, where it lays its eggs in water-filled tree holes? A: The poison dart frog.

Q: Which amphibian can change its skin texture from smooth to bumpy to match its environment? A: The gray tree frog.

Q: What amphibian species is known to "sing" during rainfall? A: The desert spadefoot toad.

Q: Which amphibian can survive in desert environments by burrowing underground during dry seasons? A: The spadefoot toad.

Q: What amphibian has developed a form of camouflage that makes it nearly invisible on moss-covered rocks? A: The Vietnamese mossy frog.

Q: Which amphibian is known for its distinctive "laughing" call? A: The laughing frog.

Q: What amphibian's skin secretes toxins so potent that even handling it without gloves can cause irritation? A: The golden poison dart frog.

Q: Which amphibian lays its eggs on land, and when the eggs hatch, the larvae wriggle into nearby water bodies? A: The salamander.

Q: What amphibian is known for its vibrant red eyes and bright green body, making it highly recognizable? A: The red-eyed tree frog.

Q: Which amphibian can endure dehydration and later rehydrate without any damage to its organs? A: The African bullfrog.

Q: What amphibian uses its long hind legs to dig burrows where it remains for months during dry periods? A: The Australian burrowing frog.

Q: Which amphibian carries its eggs in pouches on its back until they hatch into tadpoles? A: The marsupial frog.

Q: What amphibian secretes a sticky adhesive from its skin to help it cling to wet or vertical surfaces? A: The tree frog.

Q: Which amphibian can completely regenerate its tail, legs, and other body parts if lost to a predator? A: The salamander.

Q: What amphibian is known for its ability to remain underwater for long periods by absorbing oxygen through its skin? A: The African dwarf frog.

Q: Which amphibian's tongue is attached to the front of its mouth instead of the back, allowing it to quickly flick its tongue outward? A: The frog.

Q: What amphibian has been found to be able to communicate with each other using ultrasonic sounds? A: The Chinese concave-eared torrent frog.

Q: Which amphibian species hibernates during winter months by burying itself in mud? A: The common toad.

Q: What amphibian's vocal sac inflates during mating calls, allowing it to amplify sound? A: The American bullfrog.

Q: Which amphibian can leap out of the water and onto land using its powerful hind legs? A: The leopard frog.

Q: What amphibian can enter a state of brumation, a form of dormancy, during cold weather? A: The tiger salamander.

Q: Which amphibian secretes a toxic, milky substance when threatened? A: The cane toad.

Q: What amphibian has a highly developed sense of hearing, even able to detect low-frequency sounds? A: The gray tree frog.

Q: Which amphibian can use its vocalizations to establish territory during the breeding season? A: The green frog.

Q: What amphibian has been observed engaging in "anting" behavior by rubbing ants on its skin to deter parasites? A: The fire-bellied toad.

Q: Which amphibian produces a bright flash of color on its belly when disturbed to startle predators? A: The fire salamander.

Q: What amphibian's tadpoles are known to live in bromeliads, where they feed on insects and algae? A: The dart frog.

Q: Which amphibian has a "floating" stage during which its eggs float on the surface of the water? A: The African bullfrog.

Q: What amphibian can lay its eggs in foam nests suspended over water? A: The túngara frog.

Q: Which amphibian produces a toxin strong enough to deter large predators like snakes and birds? A: The rough-skinned newt.

Q: What amphibian can survive extreme heat by burrowing deep into the ground during droughts? A: The spadefoot toad.

Q: Which amphibian is known for using vocalizations to find mates over long distances in dense forests? A: The red-eyed tree frog.

Q: What amphibian species can inflate its body to twice its normal size to deter predators? A: The African bullfrog.

Q: Which amphibian can breathe underwater using gills throughout its entire life? A: The axolotl.

Q: What amphibian produces a sticky glue-like substance to secure its eggs to submerged plants? A: The common newt.

Q: Which amphibian is known for secreting a sticky, toxic foam as a defense mechanism? A: The African clawed frog.

Q: What amphibian has a symbiotic relationship with certain types of algae, which grow inside its cells and help with photosynthesis? A: The spotted salamander.

Q: Which amphibian can survive without food for up to two years by slowing down its metabolism? A: The tiger salamander.

Q: What amphibian's tadpoles are known for their cannibalistic tendencies, consuming other tadpoles in overcrowded environments? A: The spadefoot toad.

Q: Which amphibian is known for performing an elaborate "dance" during its mating ritual? A: The European common frog.

Q: What amphibian can regenerate damaged parts of its brain and spinal cord? A: The axolotl.

Q: Which amphibian has suction pads on its toes that allow it to climb vertical surfaces, including glass? A: The tree frog.

Q: What amphibian produces a poison that is toxic enough to kill several humans with just one gram? A: The golden poison dart frog.

Q: Which amphibian is known for laying its eggs in shallow water, often in large communal clusters? A: The American toad.

Q: What amphibian has a specialized vocal sac that allows it to project its call over long distances? A: The northern leopard frog.

Q: Which amphibian can lay eggs on land, which hatch into tadpoles that then crawl into the water? A: The four-toed salamander.

Q: What amphibian's bright coloration is a warning signal to predators that it is toxic? A: The poison dart frog.

Q: Which amphibian can inflate its body and produce a hissing sound when threatened by predators? A: The fire-bellied toad.

Q: What amphibian has a sticky secretion that can be used to deter predators or parasites? A: The European newt.

Q: Which amphibian is known for its remarkable ability to regenerate its limbs multiple times without scarring? A: The axolotl.

Q: What amphibian's tadpoles can survive in water bodies that dry out, by rapidly completing metamorphosis? A: The spadefoot toad.

Q: Which amphibian can survive in both fresh and brackish water environments? A: The green frog.

Q: What amphibian has a unique spiral-shaped tongue that it uses to catch insects with precision? A: The horned frog.

Q: Which amphibian produces a distinct odor from glands on its skin to deter predators? A: The common toad.

Q: What amphibian is known for creating a "foam nest" in which its eggs develop, providing protection from predators and environmental factors? A: The túngara frog.

Q: Which amphibian produces a "chorus" of sounds during the breeding season, especially after rainfall? A: The spring peeper.

Q: What amphibian can generate toxic alkaloids from its diet of ants and beetles? A: The poison dart frog.

Q: Which amphibian can secrete chemicals that inhibit the growth of bacteria and fungi, protecting its skin? A: The common newt.

Q: What amphibian's tadpoles have specialized mouthparts for scraping algae off rocks and plants? A: The African clawed frog.

Q: Which amphibian has a light-sensitive "third eye" on the top of its head to help regulate circadian rhythms? A: The spadefoot toad.

Q: What amphibian produces bright red or orange markings on its belly as a warning signal to predators? A: The fire-bellied toad.

Q: Which amphibian can reproduce in water bodies with low oxygen levels, thanks to its ability to absorb oxygen through its skin? A: The African dwarf frog.

Q: What amphibian can enter a state of torpor during extreme environmental conditions, such as drought or freezing temperatures? A: The American toad.

Q: Which amphibian's eggs are known for developing in small, temporary pools of rainwater? A: The spadefoot toad.

Q: What amphibian has a diet consisting mostly of small invertebrates, such as insects, spiders, and worms? A: The salamander.

Q: Which amphibian has a vocal sac that expands like a balloon during mating calls to increase sound projection? A: The American bullfrog.

Q: What amphibian can regenerate not only limbs but also parts of its heart and spinal cord? A: The axolotl.

Q: Which amphibian lays its eggs in large, jelly-like masses that float on the surface of ponds or lakes? A: The common frog.

Q: What amphibian has suction-cup-like toes that allow it to cling to smooth surfaces like leaves and glass? A: The tree frog.

Q: Which amphibian can adjust its body's temperature by seeking out specific microclimates, such as warm sunlit areas or cool shade? A: The salamander.

Q: What amphibian is known for laying its eggs in small, temporary water bodies that form after rainstorms? A: The spadefoot toad.

Okay, we've heard a lot about frogs and toads in the Q&A above. What the heck is the difference between them?

Frogs and toads are both amphibians, but they differ in several key ways. Frogs generally have smooth, moist skin and prefer living near water. Their long, powerful hind legs are built for jumping, and they have webbed feet that help them swim. Frogs are often more slender in shape and have bulging eyes.

Toads, on the other hand, usually have dry, warty skin and are better adapted to living on land, though they still need water to reproduce. Their legs are shorter, making them better suited for walking or short hops rather than long jumps. Toads typically have more stout, robust bodies and their eyes are less prominent than those of frogs.

Behaviorally, frogs tend to stay closer to water sources, while toads can venture farther into drier areas. Frogs lay their eggs in clusters, whereas toads lay theirs in long strings. Both are beneficial to ecosystems, as they help control insect populations, but frogs are often more associated with wet environments, and toads with dry, terrestrial habitats.

While we're on the subject of random fun facts, what in the world is a duck billed platypus and how is it classified?

The duck-billed platypus is one of the most unique animals in the world, often described as a "mammal with bird-like features." It's a semi-aquatic mammal native to Australia, known for its distinctive flat, duck-like bill, webbed feet, and beaver-like tail. The platypus is one of only five species of monotremes, a group of egg-laying mammals, which also includes the echidna.

Despite laying eggs, the platypus is classified as a mammal because it produces milk to feed its young and has fur. It also has some truly unusual traits for a mammal, such as venomous spurs on its hind legs in males and electroreception in its bill, which helps it detect prey in murky water. The combination of traits makes the platypus an evolutionary curiosity.

Scientifically, the platypus is classified as follows:

- Kingdom: Animalia
- Phylum: Chordata
- Class: Mammalia
- Order: Monotremata
- Family: Ornithorhynchidae
- Genus: Ornithorhynchus
- Species: Ornithorhynchus anatinus

Its classification reflects its mix of mammalian and reptilian traits, making the duck-billed platypus one of the most distinctive animals on Earth!

Advanced Fun Facts

Q: Which amphibian species is known for giving birth to live young rather than laying eggs? A: The alpine salamander (Salamandra atra).

Q: What is the primary purpose of the granular glands in amphibian skin? A: To secrete toxins for defense.

Q: Which amphibian can absorb oxygen directly through its skin and live without lungs for its entire life? A: The olm (Proteus anguinus).

Q: What is the evolutionary significance of amphibians having both aquatic and terrestrial life stages? A: It demonstrates their transition from water to land during vertebrate evolution.

Q: Which amphibian uses specialized mating calls that are species-specific, ensuring they attract the right mate? A: The American toad (Anaxyrus americanus).

Q: What hormone triggers metamorphosis from tadpole to adult in amphibians? A: Thyroxine (a thyroid hormone).

Q: Which amphibian exhibits neoteny, retaining larval features into adulthood? A: The axolotl (Ambystoma mexicanum).

Q: How do amphibians control water loss when they are on land? A: By regulating their skin permeability and seeking moist environments.

Q: What unique reproductive behavior is exhibited by the male Darwin's frog (Rhinoderma darwinii)? A: The male carries the developing tadpoles in his vocal sac until they are fully developed.

Q: Which amphibian is known for its ability to store urea in its body to survive in salty environments? A: The crab-eating frog (Fejervarya cancrivora).

Q: What is the primary function of amphibian vocal sacs in males? A: To amplify mating calls.

Q: Which amphibian has a specialized "mental gland" that releases pheromones to attract mates? A: The red-legged salamander (Plethodon shermani).

Q: How do amphibians detect changes in the environment using their lateral line system? A: They detect vibrations and water movements through specialized sensory organs.

Q: Which amphibian uses its sticky tongue to capture prey in less than 0.07 seconds, faster than the blink of an eye? A: The chameleon caecilian (Chthonerpeton indistinctum).

Q: What role do amphibians play in bioindicator studies in their habitats? A: They are sensitive to environmental changes, making them good indicators of ecosystem health.

Q: What amphibian has the longest known larval stage, taking up to 7 years to metamorphose? A: The European olm (Proteus anguinus).

Q: Which amphibian can regenerate its heart and other internal organs if damaged? A: The axolotl (Ambystoma mexicanum).

Q: How does the chytrid fungus (Batrachochytrium dendrobatidis) affect amphibians? A: It disrupts their skin's ability to absorb water and electrolytes, leading to death.

Q: What amphibian exhibits "explosive breeding," where large numbers of individuals breed over a very short period? A: The wood frog (Lithobates sylvaticus).

Q: Which amphibian has a "barber pole" pattern inside its tongue that helps catch prey with precision? A: The horned frog (Ceratophrys).

Q: What type of toxin do fire-bellied toads secrete that makes them distasteful to predators? A: Bombesin.

Q: Which amphibian has a specialized "vomeronasal organ" to detect pheromones during mating? A: The tiger salamander (Ambystoma tigrinum).

Q: What amphibian species has external fertilization, where the male deposits sperm over the female's eggs in water? A: The African clawed frog (Xenopus laevis).

Q: How does the glass frog's translucent skin benefit it in its environment? A: It provides camouflage by blending in with its surroundings and reducing visibility to predators.

Q: Which amphibian is capable of enduring high levels of radiation without suffering harm? A: The alpine newt (Ichthyosaura alpestris).

Q: What unique parental care strategy is used by the male midwife toad (Alytes obstetricans)? A: The male carries fertilized eggs wrapped around his hind legs until they hatch.

Q: Which amphibian uses a specialized "courtship dance" involving body undulations to attract mates? A: The crested newt (Triturus cristatus).

Q: What specialized sensory system allows salamanders to detect chemical cues from prey in water? A: The olfactory system combined with their vomeronasal organ.

Q: Which amphibian species has evolved aposematic coloration to warn predators of its toxicity? A: The red-spotted newt (Notophthalmus viridescens).

Q: How do frogs produce their loud mating calls despite their small size? A: By inflating their vocal sacs to project sound across long distances.

Q: Which amphibian uses its feet to dig backwards into the soil, forming burrows to escape the heat? A: The Mexican burrowing toad (Rhinophrynus dorsalis).

Q: What is the primary role of the "parotoid glands" in toads? A: To secrete toxins that deter predators.

Q: How do some amphibians such as the Iberian ribbed newt (Pleurodeles waltl) defend themselves when threatened? A: They push their ribs through their skin to deliver toxins to predators.

Q: Which amphibian species migrates en masse to breeding ponds, sometimes crossing roads in large numbers? A: The common toad (Bufo bufo).

Q: What amphibian uses a specialized form of predation called "sit-and-wait," where it remains motionless until prey approaches? A: The horned frog (Ceratophrys).

Q: Which amphibian has evolved a flattened body to hide in tight crevices during the dry season? A: The hellbender salamander (Cryptobranchus alleganiensis).

Q: What amphibian can produce a unique, sticky mucus that aids in climbing slippery surfaces? A: The European tree frog (Hyla arborea).

Q: Which amphibian exhibits a reproductive strategy known as "egg guarding," where one parent remains with the eggs until they hatch? A: The strawberry poison dart frog (Oophaga pumilio).

Q: What amphibian species can live for extended periods in underground burrows to avoid drought conditions? A: The spadefoot toad (Scaphiopus holbrookii).

Q: Which amphibian uses "biofluorescence," where it absorbs light and re-emits it as a different color, to communicate or attract mates? A: The polka-dot tree frog (Boana punctata).

Q: How do some amphibians avoid detection by predators using their "startle reflex"? A: They display bright, hidden colors suddenly to startle predators.

Q: Which amphibian exhibits polyandry, where a female mates with multiple males during the breeding season? A: The common toad (Bufo bufo).

Q: What is the main environmental factor that triggers amphibian breeding behaviors? A: Rainfall and increased humidity.

Q: Which amphibian species exhibits territorial behavior, defending breeding sites from rivals? A: The poison dart frog (Dendrobates spp.).

Q: How do amphibians detect minute chemical changes in water that signal the presence of predators? A: Through specialized chemoreceptors in their skin.

Q: What adaptation allows amphibians like the mudpuppy (Necturus maculosus) to remain aquatic throughout their lives? A: External gills and a lack of metamorphosis.

Q: Which amphibian can produce chemicals in its skin that are being researched for potential medical uses, such as pain relief or antibiotics? A: The poison dart frog (Phyllobates terribilis).

Q: How do some species of frogs use "foam nests" to protect their eggs from predators and environmental threats? A: They create a frothy mixture that surrounds the eggs and keeps them moist.

Q: Which amphibian is capable of aestivation, a dormancy period during hot or dry seasons? A: The African bullfrog (Pyxicephalus adspersus).

Q: What evolutionary adaptation allows amphibians to avoid predators through mimicry? A: Some amphibians, like the mimic poison frog (Ranitomeya imitator), mimic the bright colors of toxic species to deter predators.

Finally, here's a little Latin challenge for you with a list of 20 popular scientific names for amphibians.

Rana temporaria – Common Frog
Bufo bufo – Common Toad
Hyla cinerea – American Green Tree Frog
Ambystoma mexicanum – Axolotl
Lithobates catesbeianus – American Bullfrog
Pleurodeles waltl – Iberian Ribbed Newt
Notophthalmus viridescens – Eastern Newt (Red-Spotted Newt)
Alytes obstetricans – Common Midwife Toad
Xenopus laevis – African Clawed Frog
Bombina orientalis – Oriental Fire-Bellied Toad
Agalychnis callidryas – Red-Eyed Tree Frog
Ambystoma tigrinum – Tiger Salamander
Litoria caerulea – Australian Green Tree Frog
Rhinoderma darwinii – Darwin's Frog
Pseudacris crucifer – Spring Peeper

Ceratophrys ornata – Ornate Horned Frog (Pacman Frog)
Dendrobates tinctorius – Dyeing Poison Dart Frog
Triturus cristatus – Great Crested Newt
Scaphiopus holbrookii – Eastern Spadefoot Toad
Hymenochirus boettgeri – African Dwarf Frog

Have you ever heard of Amazon jungle natives using poison darts made by rubbing their arrow tips on the backs of frogs? While used primarily to stun or paralyze prey, Hollywood suggests the poison darts and arrows are also effective in battle.

Maybe. Anyway, here are some details on frog poisoned arrows.

In the Amazon jungle, indigenous tribes have long practiced the art of creating poison darts or arrows, using a deadly toxin derived from certain species of frogs, most notably poison dart frogs (Dendrobatidae family). This traditional practice is both a survival tool and an integral part of the culture and hunting methods in the region.

Poison dart frogs, particularly species like the golden poison frog (Phyllobates terribilis), secrete highly potent toxins from their skin. These toxins, called batrachotoxins, are some of the most powerful natural substances, capable of disrupting nerve and muscle functions, leading to paralysis and death. Interestingly, the frogs themselves are not born with these toxins. Instead, they accumulate them by consuming certain insects, such as ants and mites, which are found in their native environment.

To make the poison for darts or arrows, indigenous hunters carefully capture the frogs without directly handling them, as even touching the frogs can be dangerous. The frogs are then gently heated over a fire, which stimulates the secretion of the toxic chemicals through their skin. The hunters collect this secretion and dip their blowgun darts or arrow tips into the poison.

The darts or arrows used in hunting are typically small and light, made from materials like sharpened bamboo or wood. Once the tips are coated in the frog's poison, they are highly lethal. These darts are often launched from blowguns, a silent and effective tool for hunting in dense jungle environments. The hunters aim for small prey like birds, monkeys, and other small animals, and the poison quickly immobilizes or kills the prey without the need for a direct hit to vital organs.

While the poison is lethal, it breaks down in heat, meaning that the meat of animals hunted with these darts remains safe to eat after cooking. The tradition of making poison darts is passed down through generations, along with the knowledge of which frogs are the most toxic and how to handle them safely. This practice showcases the deep connection between indigenous peoples and their environment, as they understand and harness the natural resources of the Amazon in a sustainable manner.

This method of hunting has been used for centuries and remains an important aspect of Amazonian tribal culture. The process is a prime example of how indigenous knowledge is deeply tied to the biodiversity of the rainforest and the specific ecosystems in which these communities thrive.

Fish

Fish are cold-blooded animals that live in water and are known for their ability to breathe through gills. Unlike mammals, fish don't need lungs; they get oxygen by filtering water through their gills. Most fish have fins to help them swim and a streamlined body that allows them to move easily through the water.

Fish are covered in scales that protect their bodies, and many have a slimy coating that reduces friction in the water and helps them avoid parasites. Fish come in many shapes and sizes, from tiny goldfish to massive sharks.

There are three main types of fish:

1. Bony fish (like salmon and trout), which have skeletons made of bone.
2. Cartilaginous fish (like sharks and rays), which have skeletons made of cartilage, a softer, more flexible tissue.
3. Jawless fish (like lampreys), which are much simpler and lack jaws.

Fish live in a wide variety of environments, from freshwater rivers and lakes to salty oceans and seas. They are an important part of the food chain, with some fish eating plants or smaller animals and others being predators at the top of their ecosystem.

Fish reproduce by laying eggs, though some species, like certain sharks, give birth to live young. They are a vital resource for humans, providing food and supporting ecosystems worldwide.

Q: What is the fastest fish in the ocean? A: The sailfish, reaching speeds up to 68 mph.

Q: Which fish can live out of water for several days by breathing air through its skin and mouth? A: The walking catfish.

Q: What is the largest species of fish? A: The whale shark, growing up to 40 feet long.

Q: Which fish has the ability to generate electric shocks up to 600 volts? A: The electric eel.

Q: What fish can inflate its body to avoid predators? A: The pufferfish.

Q: Which fish has the ability to swim backward? A: The knifefish.

Q: What is the longest-living fish species? A: The Greenland shark, with some living over 400 years.

Q: Which fish is known for its incredible leaping abilities, sometimes jumping over 10 feet out of water? A: The Asian carp.

Q: What fish can change its sex during its lifetime, often switching from female to male? A: The clownfish.

Q: Which fish has scales that reflect polarized light, making it nearly invisible to predators in the water? A: The Atlantic herring.

Speaking of the Atlantic, one of my favorite fish is the Atlantic Cod. The Atlantic cod has long been one of the most important fish in the world, especially in terms of commercial value. This fish was a staple for centuries, with entire economies built around its catch and trade. From the New England coast to Canada and Europe, cod was everywhere. It's known for its mild flavor and flaky texture, making it a favorite for dishes like fish and chips.

But here's the problem: overfishing. Cod populations were once so massive that people thought they'd never run out. Fishing boats would bring in huge hauls daily, and technology only made it easier to catch more. Unfortunately, no one really paid attention to the signs that cod stocks were getting dangerously low. By the late 20th century, cod populations in the North Atlantic had collapsed.

Governments scrambled to impose fishing limits, but for some regions, like the Grand Banks near Canada, the damage was already done. Cod were nearly driven to extinction in certain areas, and entire fishing communities suffered as their livelihoods disappeared.

Q: What fish has a "second set of jaws" known as pharyngeal jaws to help it crush food? A: The moray eel.

Q: Which fish can use bioluminescence to attract prey in deep ocean environments? A: The anglerfish.

Q: What fish has a transparent head, allowing it to see through its skull? A: The barreleye fish.

Q: Which fish is capable of walking on land using its pectoral fins? A: The mudskipper.

Q: What fish can detect electrical fields produced by other animals to locate prey? A: The hammerhead shark.

The great oceanographer and film maker Jacques Cousteau once stated, "When man enters the water, he enters the food chain…and not necessarily at the top." If you've ever encountered a shark in the water, you know how true those words are.

Human fatalities due to shark attacks are actually quite rare, considering how many people swim in the ocean every day. Despite sharks having a fearsome reputation, the number of shark-related deaths worldwide is typically low, averaging about 5 to 10 per year. Most shark species aren't interested in humans as prey, and many attacks are thought to be cases of mistaken identity, where sharks bite surfers or swimmers, mistaking them for seals or other prey.

That said, there have been some historically bad years for shark attacks. 1916 stands out as one of the worst, with the infamous Jersey Shore shark attacks in the United States. Over the span of just 12 days, five people were attacked, and four of them died. These attacks happened in a time when people were not as aware of sharks as a danger, and they occurred in unexpected locations, like a freshwater creek. The attacks sparked widespread fear and hysteria, later inspiring the novel and movie *Jaws*.

Another notably bad year was 2020, which saw an uptick in shark fatalities, with 10 deaths globally. While shark attacks capture the public's imagination, it's important to remember that fatalities are extremely rare, and sharks are more threatened by humans than the other way around.

Q: Which fish can survive being frozen solid and then thawed out without damage? A: The Antarctic toothfish.

Q: What fish has the ability to camouflage itself by changing its color to blend into its surroundings? A: The flounder.

Q: Which fish has venomous spines that can cause severe pain to humans? A: The lionfish.

Q: What fish is known for its ability to glow in the dark due to bioluminescent bacteria? A: The flashlight fish.

Q: Which fish can produce a slime that clogs the gills of predators, deterring attacks? A: The hagfish.

Q: What fish has been known to live at depths exceeding 26,000 feet in the ocean? A: The snailfish.

Q: Which fish is known for building elaborate sand structures to attract mates? A: The pufferfish (specifically, the white-spotted puffer).

Q: What fish has the ability to "talk" by producing sound through its swim bladder? A: The croaker fish.

Q: Which fish can deliver a painful, venomous sting through its spiny dorsal fin? A: The stonefish.

Q: What fish species can live in both freshwater and saltwater environments, transitioning between the two? A: The salmon.

Q: Which fish is capable of producing toxic mucus to defend itself from predators? A: The parrotfish.

Q: What fish has the ability to regenerate lost fins and other damaged body parts? A: The zebrafish.

Q: Which fish is known for its incredible speed and hunting prowess, often catching prey in mid-air? A: The blue marlin.

Q: What fish has suction-cup-like fins that allow it to cling to rocks in fast-moving streams? A: The clingfish.

Q: Which fish has the unique ability to produce "milk" to feed its young after birth? A: The discus fish.

Q: What fish has the most elaborate courtship dance, involving body color changes and fin displays? A: The mandarin fish.

Q: Which fish can breathe air directly from the atmosphere using a modified swim bladder? A: The lungfish.

Q: What fish has the ability to swim faster than a cheetah can run? A: The sailfish.

When it comes to the title of "ugliest fish in the world," the blobfish usually wins by a landslide! This deep-sea fish, found off the coasts of Australia and New Zealand, looks like a blob of gelatinous goo when it's out of water, with a droopy face, sagging body, and a kind of grumpy, melted appearance. But here's the thing: blobfish only look this way when they're brought to the surface.

In their natural habitat, miles below the ocean's surface, the blobfish is under extreme pressure, which keeps its body looking much more normal. Since they live in such deep waters, they don't need muscles or bones like other fish; their squishy bodies help them float just above the ocean floor, where they eat whatever drifts by.

So, while they might look hilariously ugly to us on land, the blobfish is perfectly adapted for its deep-sea life!

Q: Which fish species is known for carrying its eggs in its mouth for protection? A: The mouthbrooding cichlid.

Q: What fish species has scales so strong they have been used by indigenous peoples to make tools and armor? A: The arapaima.

Q: Which fish species is hermaphroditic, possessing both male and female reproductive organs? A: The hamlet fish.

Q: What fish is known for leaping out of the water to escape predators and sometimes landing in boats? A: The flying fish.

Q: Which fish can see ultraviolet light, helping it detect prey and navigate? A: The bluefin tuna.

Q: What fish species has the largest eyes relative to its body size? A: The bigeye thresher shark.

Q: Which fish is capable of emitting an electrical discharge to stun prey or defend itself from predators? A: The electric ray.

Q: What fish has the ability to "play dead" to avoid being eaten by predators? A: The gar.

Q: Which fish has a special gland that produces antifreeze proteins, preventing its blood from freezing in icy waters? A: The Antarctic cod.

Q: What fish can burrow into the mud and survive months of drought by entering a state of dormancy? A: The African lungfish.

Q: Which fish can inflate its body to appear larger and more intimidating to predators? A: The porcupinefish.

Q: What fish has one of the strongest bites in the animal kingdom, capable of crushing shells and coral? A: The black piranha.

Q: Which fish can communicate with others of its species using a series of grunts, clicks, and pops? A: The plainfin midshipman.

Q: What fish can walk on the ocean floor using its modified fins? A: The frogfish.

Q: Which fish is known for its highly territorial behavior, defending its space aggressively? A: The betta fish.

Q: What fish can survive in water with almost no oxygen by gulping air from the surface? A: The labyrinth fish.

Q: Which fish can produce a paralyzing toxin that deters predators, even larger fish? A: The boxfish.

Q: What fish species has eyes that can rotate independently of each other, giving it a 360-degree view? A: The chameleon fish.

Q: Which fish has a symbiotic relationship with cleaner shrimp, allowing the shrimp to clean its scales and gills? A: The grouper.

Q: What fish species is known for building nests in gravel or sand for laying its eggs? A: The stickleback.

Q: Which fish can deliver a venomous sting through its fins, causing extreme pain to humans? A: The scorpionfish.

Q: What fish species is famous for changing color during courtship displays or when threatened? A: The cuttlefish.

Q: Which fish can leap out of the water to catch insects or small animals on overhanging branches? A: The archerfish.

Q: What fish has the ability to swim in open ocean waters and also in rivers and freshwater lakes? A: The bull shark.

Q: Which fish has been found to use tools, such as rocks, to crack open shellfish? A: The tuskfish.

Q: What fish can blend into its environment by mimicking the appearance of other marine creatures, like seaweed? A: The leafy sea dragon.

Q: Which fish has the ability to glow in the dark due to natural bioluminescence? A: The lanternfish.

Q: What fish has the largest known migration of any marine species, traveling thousands of miles across oceans? A: The Pacific bluefin tuna.

Q: Which fish species can switch its reproductive role multiple times throughout its life? A: The parrotfish.

Q: What fish has a sucker-like mouth that allows it to attach to larger marine animals, like sharks and whales? A: The remora.

Q: Which fish has evolved to live at extreme depths, where the pressure would crush most other species? A: The fangtooth.

Q: What fish is known for traveling great distances upstream to spawn in freshwater, only to die afterward? A: The salmon.

Q: Which fish uses a modified spine to create a powerful "punch," sometimes breaking aquarium glass? A: The mantis shrimp (though technically a crustacean, its behavior is fish-like in nature).

Q: What fish species produces slime as a defense mechanism, sometimes clogging the gills of predators? A: The hagfish.

Q: Which fish has the ability to generate electrical fields for communication, navigation, and hunting? A: The knifefish.

Q: What fish has one of the strongest jaws in the ocean, allowing it to crush hard shells with ease? A: The triggerfish.

Q: Which fish can "hear" through vibrations detected in its lateral line system? A: The catfish.

Q: What fish species has been known to use "tactical retreats," hiding in sand or coral to ambush prey? A: The stargazer fish.

Q: Which fish can live in brackish water and is often found in mangrove swamps? A: The molly.

Q: What fish has a series of spines along its body that it can erect when threatened by predators? A: The pufferfish.

Q: Which fish can survive in water with very low oxygen levels by using specialized gills? A: The gourami.

Q: What fish has a suction disc on its body that allows it to stick to smooth surfaces like rocks or glass? A: The hillstream loach.

Q: Which fish is known for having one of the longest migrations of any freshwater species? A: The European eel.

Q: What fish has an extendable jaw that allows it to catch prey at lightning speed? A: The goblin shark.

Q: Which fish species can switch between "sleep" and "wake" states without closing its eyes? A: The parrotfish.

Q: What fish uses water jets to dig into the sand for food and protection? A: The garden eel.

Q: Which fish has a venomous bite, rather than venomous spines, to deliver toxins to prey? A: The fangtooth moray.

Q: What fish has scales that act as armor, protecting it from predators in its riverine habitat? A: The alligator gar.

Q: Which fish has a fused jaw that allows it to crush coral and other hard substrates? A: The parrotfish.

Q: What fish has a specially adapted swim bladder that allows it to "hover" at different depths in the water? A: The grouper.

Q: Which fish species is known for building small rock piles as part of its courtship behavior? A: The cichlid.

Q: What fish has one of the longest lifespans of any marine species, sometimes living over 100 years? A: The orange roughy.

Q: Which fish is known for having the ability to absorb its tail into its body if damaged? A: The ribbon eel.

Q: What fish has a specialized jaw that allows it to create a "vacuum" to suck in prey? A: The bass.

Q: Which fish can "play dead" when it senses a predator approaching, remaining motionless on the ocean floor? A: The flounder.

Q: What fish has a venomous spine located on its tail, used primarily for defense? A: The stingray.

Q: Which fish can communicate through rapid color changes that signal social status or mood? A: The wrasse.

Q: What fish can survive in temperatures close to freezing, thanks to a natural antifreeze in its blood? A: The icefish.

Q: Which fish can produce clicking or grunting sounds to communicate with others of its species? A: The grunt fish.

Q: What fish has a long, thin body and can swim with incredible agility in coral reefs? A: The trumpetfish.

Q: Which fish is known for its ability to travel upstream in rivers and over obstacles during migration? A: The trout.

Q: What fish has the ability to climb waterfalls and other vertical surfaces using its suction-cup-like fins? A: The goby.

Q: Which fish species is known for creating complex tunnels and hiding places in sand or gravel? A: The jawfish.

Q: What fish can detect the Earth's magnetic field, using it to navigate during long migrations? A: The salmon.

Q: Which fish has the ability to "mimic" the appearance of other marine creatures, including venomous species, to avoid predators? A: The mimic octopus (though technically an octopus, it shares fish-like behavior).

Q: What fish has the ability to release a cloud of ink-like substance when threatened, similar to cephalopods? A: The cuttlefish (again, more of a cephalopod, but shares behaviors with fish).

Q: Which fish has scales that function as armor, providing defense against predators in rivers and lakes? A: The armored catfish.

Human deaths caused by interactions with fish are relatively rare compared to other animals, but they do happen under specific circumstances. These deaths usually occur due to accidents, poisonous species, or dangerous encounters with large marine creatures. Below are some key statistics and common causes of fish-related fatalities:

1. Shark Attacks

Global Shark Attacks: While sharks are the most feared fish, fatal shark attacks are exceedingly rare. According to the International Shark Attack File (ISAF), there are about 80 unprovoked shark attacks annually worldwide, with an average of 5 to 10 resulting in fatalities.

Fatal Shark Attacks (2022): In 2022, ISAF recorded 5 fatal unprovoked shark attacks globally, with most occurring in Australia, the U.S., and South Africa.

2. Stingrays

Stingray Incidents: Fatal stingray encounters, like the one that tragically killed Steve Irwin, are extremely rare. Most stingray injuries result from defensive stings when humans accidentally step on them.

Fatality Rate: The global fatality rate from stingray attacks is incredibly low, with only a few documented deaths, most resulting from stings to the chest or abdomen.

3. Poisonous Fish

Stonefish: The stonefish (Synanceia) is one of the most venomous fish in the world. Its venom can cause extreme pain, paralysis, and even death if not treated. However, fatalities are rare due to the availability of antivenom.

Pufferfish (Fugu): Pufferfish contain tetrodotoxin, a potent neurotoxin. In Japan, deaths occur from improper preparation of fugu, a delicacy made from pufferfish. On average, about 2 to 5 deaths per year are reported from fugu poisoning.

Lionfish: While not typically fatal, lionfish stings can cause extreme pain and allergic reactions. Fatal incidents are exceedingly rare, but the venom can be dangerous to vulnerable individuals.

4. Electric Fish

Electric Eels: Although deaths from electric eels (Electrophorus electricus) are rare, these fish can deliver shocks of up to 600 volts. While most people recover, fatalities have occurred due to drowning or heart failure after multiple shocks.

5. Fishing-Related Accidents

Fishing Fatalities: Commercial and recreational fishing accidents are more common causes of death related to fish. According to the U.S. Coast Guard, between 2000 and 2019, commercial fishing fatalities accounted for 1,201 deaths in the U.S. alone, primarily due to drowning, falls overboard, and vessel accidents.

6. Jellyfish (Though Not Fish)

Box Jellyfish: While jellyfish are not technically fish, box jellyfish stings cause several fatalities each year, primarily in Australia and Southeast Asia. The venom can cause cardiac arrest, and death can occur within minutes if untreated.

Overall, human deaths from interactions with fish are rare, especially compared to land animals or other marine species. Shark attacks garner the most media attention but are infrequent. Poisonous fish, like stonefish and pufferfish, can cause fatalities, but advances in treatment and safety precautions have minimized their impact. Most fishing-related fatalities come from accidents rather than interactions with dangerous species.

Most fish, other than sharks, are not aggressive towards humans. However, in a sad case, Steve Irwin, the famous Australian wildlife expert and conservationist, was tragically killed by a stingray. Specifically, it was a short-tail stingray (Dasyatis brevicaudata) while filming a documentary in the Great Barrier Reef in 2006. The stingray's barb pierced his chest, causing fatal injuries. Stingrays generally aren't aggressive, but they can defend themselves with a venomous spine on their tail if they feel threatened. In this rare and unfortunate case, the stingray's barb punctured Irwin's heart, leading to his death.

Advanced Fun Facts

Q: What is the scientific term for a fish's ability to regulate its buoyancy in the water column? A: The swim bladder.

Q: Which fish is considered the fastest swimmer in the world, reaching speeds of up to 68 mph? A: The sailfish (Istiophorus platypterus).

Q: What fish has the largest brain-to-body ratio of any vertebrate? A: The elephantnose fish (Gnathonemus petersii).

Q: Which fish species is known to produce a weak electric field to navigate and communicate? A: The electric knifefish (Gymnotiformes).

Q: What is the deepest-dwelling fish species ever recorded? A: The Mariana snailfish (Pseudoliparis swirei), found at depths of over 26,000 feet.

Q: Which fish uses its unique protrusible jaw to suck prey into its mouth rapidly? A: The slingjaw wrasse (Epibulus insidiator).

Q: What type of fish produces venom in its spines and can cause severe pain, sometimes death? A: The stonefish (Synanceia).

Q: Which species of fish can survive out of water for months by breathing through its skin and air-breathing organs? A: The African lungfish (Protopterus).

Q: What is the term for fish species that are born in freshwater, migrate to saltwater to mature, and then return to freshwater to spawn? A: Anadromous fish (e.g., salmon).

Q: Which fish can generate electrical charges of up to 600 volts to stun prey or defend itself? A: The electric eel (Electrophorus electricus).

Q: What fish has a specialized structure known as the "Weberian apparatus," which enhances its hearing abilities? A: The catfish.

Q: Which fish produces a cloud of bioluminescent mucus to escape from predators? A: The ponyfish (Leiognathidae).

Q: What fish has a unique antifreeze protein in its blood, allowing it to survive in near-freezing temperatures? A: The Antarctic toothfish (Dissostichus mawsoni).

Q: What is the largest species of ray, capable of reaching wingspans over 20 feet? A: The manta ray (Manta birostris).

Q: Which fish is known for "sequential hermaphroditism," where it changes sex as it matures? A: The clownfish (Amphiprioninae).

Q: Which fish can shoot jets of water at insects above the surface to knock them into the water for feeding? A: The archerfish (Toxotes jaculatrix).

Q: What fish has evolved a mouth structure that allows it to filter feed plankton efficiently while swimming with its mouth open? A: The whale shark (Rhincodon typus).

Q: What is the longest recorded migration of any fish species? A: The Pacific salmon (Oncorhynchus spp.), migrating over 3,000 miles to spawn.

Q: Which fish species has a modified pelvic fin that allows it to attach itself to other fish or objects? A: The remora (Echeneidae).

Q: What fish can inflate its body by gulping air or water, making it more difficult for predators to swallow? A: The pufferfish (Tetraodontidae).

Q: Which fish has an organ called a "spiral valve," used to increase surface area for nutrient absorption in digestion? A: The shark.

Q: What fish species is capable of regenerating its heart and other tissues after damage? A: The zebrafish (Danio rerio).

Q: Which fish produces slime that expands in water to suffocate predators, and is known for producing copious amounts of this slime? A: The hagfish (Myxini).

Q: Which fish uses its "ampullae of Lorenzini" to detect electromagnetic fields generated by other animals? A: The shark.

Q: What fish species can survive in highly acidic or alkaline water environments, often toxic to other species? A: The tilapia (Oreochromis spp.).

Q: Which fish is known for having bioluminescent bacteria that help it attract prey in deep-sea environments? A: The anglerfish (Lophiiformes).

Q: What fish has the highest tolerance for saltwater and freshwater, easily transitioning between the two environments? A: The bull shark (Carcharhinus leucas).

Q: What fish is known for changing its coloration dramatically during mating displays or social interactions? A: The parrotfish (Scaridae).

Q: Which fish has the longest known lifespan, sometimes living over 200 years? A: The rougheye rockfish (Sebastes aleutianus).

Q: What fish uses a specialized light-producing organ to lure prey in the pitch-dark deep sea? A: The black dragonfish (Idiacanthus atlanticus).

Q: Which fish species has been found to have the ability to detect Earth's magnetic fields for navigation during migrations? A: The European eel (Anguilla anguilla).

Q: What fish has specialized fins that allow it to "walk" along the ocean floor? A: The frogfish (Antennariidae).

Q: Which fish can produce toxins strong enough to kill several adult humans, with no known antidote? A: The pufferfish (Tetraodontidae), which contains tetrodotoxin.

Q: What fish species is known for its complex bubble-nest building behavior during breeding? A: The betta fish (Betta splendens).

Q: Which fish has a cartilaginous skeleton rather than a bony one? A: The great white shark (Carcharodon carcharias).

Q: What fish has the unique ability to grow back damaged parts of its fins, spine, and even its eyes? A: The zebrafish (Danio rerio).

Q: Which fish uses rapid fin movements to generate water jets for propulsion and hunting? A: The cuttlefish (though not a fish, its behavior is fish-like).

Q: What fish produces milk-like secretions from its skin to nourish its young after they hatch? A: The discus fish (Symphysodon).

Q: Which fish has a unique bone structure in its head, allowing it to detect vibrations and low-frequency sounds? A: The catfish.

Q: What fish is capable of surviving in anoxic (low oxygen) environments by switching to anaerobic respiration? A: The crucian carp (Carassius carassius).

Q: Which fish is known for its "cleaner" behavior, removing parasites from larger marine animals? A: The cleaner wrasse (Labroides dimidiatus).

Q: What fish is capable of producing venom through a modified spine located near its tail or fins? A: The lionfish (Pterois).

Q: Which fish species is known for producing sounds to communicate, sometimes referred to as "vocal fish"? A: The toadfish (Opsanus tau).

Q: What fish is known for being highly territorial, even attacking its own reflection when housed in aquariums? A: The betta fish (Betta splendens).

Q: Which fish uses a specialized "bio-lens" in its eyes to focus on both nearby and distant objects in the deep sea? A: The spookfish (Dolichopteryx longipes).

Q: What fish can survive freezing temperatures by producing antifreeze proteins that prevent ice crystals from forming in its blood? A: The Antarctic cod (Dissostichus mawsoni).

Q: Which fish is known for producing the longest electrical pulses to detect objects in murky waters? A: The elephantnose fish (Gnathonemus petersii).

Q: What fish can see in near-total darkness, thanks to a specialized adaptation in its retinas? A: The bigeye fish (Priacanthidae).

Q: Which fish species is famous for undergoing a complete physical and color transformation when switching sex? A: The wrasse (Labridae).

Q: What fish is capable of producing electric discharges for hunting and self-defense, often mistaken for an eel? A: The electric catfish (Malapteruridae).

Finally, here are some scientific names for fish you may recognize.

Betta splendens – Betta (Siamese Fighting Fish)
Carassius auratus – Goldfish
Danio rerio – Zebrafish
Pterophyllum scalare – Angelfish
Paracheirodon innesi – Neon Tetra
Corydoras aeneus – Bronze Corydoras (Cory Catfish)
Poecilia reticulata – Guppy
Xiphophorus hellerii – Swordtail
Paracheirodon axelrodi – Cardinal Tetra
Symphysodon aequifasciatus – Discus Fish
Mollienesia latipinna – Sailfin Molly
Osteoglossum bicirrhosum – Silver Arowana
Scleropages formosus – Asian Arowana

Balantiocheilos melanopterus – Bala Shark
Astronotus ocellatus – Oscar Fish
Gyrinocheilus aymonieri – Chinese Algae Eater
Tetraodon nigroviridis – Green Spotted Pufferfish
Ctenopharyngodon idella – Grass Carp
Pygocentrus nattereri – Red-Bellied Piranha
Hypostomus plecostomus – Common Pleco

Crustaceans

Crustaceans are a group of animals that live mostly in water and are part of the larger group known as arthropods, which also includes insects and spiders. They have hard exoskeletons (external skeletons) made of a tough material called chitin that protects their bodies. As crustaceans grow, they need to molt, or shed their exoskeleton, to get bigger.

Crustaceans have bodies divided into segments, and most have jointed legs. Many crustaceans, like crabs and lobsters, have claws they use for catching food and defending themselves. They also have two pairs of antennae that help them sense their environment, which is more than most other arthropods.

There are many different types of crustaceans, including:

1. Crabs and lobsters, which are some of the larger and more well-known types.
2. Shrimp, which are smaller and often live in both freshwater and saltwater.

3. Barnacles, which attach themselves to rocks or boats and filter food from the water.
4. Isopods, like the pill bug (also known as a roly-poly), which are crustaceans that can live on land.

Crustaceans can be carnivores, herbivores, or scavengers, feeding on a wide variety of foods, including plants, other animals, and dead matter. They play an important role in marine ecosystems and are a major food source for humans and other animals.

Fun Fact: Not all crabs are crustaceans.

Crabs (the STD, not the delicious seafood) are actually tiny parasitic lice known as Pthirus pubis that set up camp in your pubic hair. These little party crashers love cozying up in coarse body hair, so they don't discriminate between the beach and your, uh, personal area. They spread mainly through close physical contact (yep, that kind), but can also hitch a ride on towels or clothing. Once settled in, they start biting, leaving you itchy and awkwardly scratching in public.

As for their place in the animal world, crabs (the bugs) belong to the order Phthiraptera, which includes all lice. So, while they might seem like crustaceans at a glance, these little jerks are definitely insects, not actual crabs. But hey, they've got the same annoying persistence as their sea-dwelling namesakes—just without the tasty meat or claws!

Moving on, here are a couple hundred random crustacean related pub quiz style question and answer trivia facts to keep you awake and reading throughout the night.

Q: What is the largest species of crustacean? A: The Japanese spider crab, with a leg span of up to 12 feet.

Q: Which crustacean can regenerate lost limbs, including claws? A: The fiddler crab.

Q: What crustacean is known for its powerful claw that snaps shut with incredible speed, creating a cavitation bubble? A: The snapping shrimp (also known as the pistol shrimp).

Q: Which crustacean can live in both freshwater and terrestrial environments? A: The land crab.

Q: What crustacean species is known for its ability to change color to match its environment? A: The rock lobster.

Q: Which crustacean carries its eggs under its abdomen until they hatch? A: The blue crab.

Q: What is the smallest species of crustacean? A: The Stygotantulus stocki, measuring only 0.1 mm in length.

Q: Which crustacean is capable of living up to 100 years in captivity? A: The American lobster.

Q: What crustacean is known for creating "burrows" in sand or mud to protect itself from predators? A: The ghost shrimp.

Q: Which crustacean can filter feed by using feather-like appendages to catch plankton? A: The barnacle.

Q: What is the only known crustacean species that can survive out of water for extended periods by retaining water in its gills? A: The coconut crab.

Q: Which crustacean species is known for its bioluminescent properties, glowing in the dark? A: The ostracod (specifically, some species of seed shrimp).

Q: What crustacean uses its claws to communicate with others in its colony through rhythmic "drumming" sounds? A: The ghost crab.

Q: Which crustacean is capable of burrowing up to 6 feet deep into the ocean floor? A: The mud crab.

Q: What crustacean is the only one known to produce sound by rubbing its legs together, similar to crickets? A: The spiny lobster.

Q: Which crustacean can travel on land and is sometimes referred to as a "land lobster"? A: The coconut crab.

Q: What crustacean's shell contains pigments that change color when boiled? A: The lobster (its shell turns red when cooked due to heat-sensitive pigments).

Q: Which crustacean uses its antennae to taste and smell its surroundings? A: The crayfish.

Q: What is the largest freshwater crustacean in the world? A: The Tasmanian giant freshwater crayfish.

Q: Which crustacean is known to molt its entire exoskeleton to grow, a process called ecdysis? A: The blue crab.

Q: What crustacean species lives in symbiosis with sea anemones, using them for protection? A: The boxer crab.

Q: Which crustacean's eyes can detect polarized light, giving it advanced visual capabilities? A: The mantis shrimp.

Q: What crustacean is known for its one enlarged claw, which it waves to attract mates? A: The fiddler crab.

Q: Which crustacean is able to survive in both saltwater and freshwater environments, even transitioning between the two? A: The Chinese mitten crab.

Q: What crustacean can leap up to 10 times its body length to escape predators? A: The sand flea.

Q: Which crustacean can live without food for up to two years by entering a state of dormancy? A: The crayfish.

Q: What crustacean can consume up to 1,000 times its body weight in food each year? A: The shrimp.

Q: Which crustacean can sense chemical signals in the water to detect potential mates or predators? A: The lobster.

Q: What crustacean species is known to carry its young on its back, similar to how some mammals carry their offspring? A: The amphipod.

Q: Which crustacean can walk forward, backward, and sideways? A: The crab.

Q: What crustacean has 10 legs, with the first pair often modified into claws (chelae)? A: The decapod (e.g., crabs, lobsters, shrimp).

Q: Which crustacean produces an iridescent shine on its shell, giving it a metallic appearance? A: The mantis shrimp.

Q: What crustacean can release a toxic substance from its body as a defense mechanism? A: The woodlouse.

Q: Which crustacean can change its exoskeleton to a darker color in response to environmental stress? A: The snow crab.

In the movie *Cast Away*, Tom Hanks' character catches a crab that oozes an unappetizing, gross liquid when cracked open. This is likely a coconut crab, which is found in tropical regions like the Pacific islands where the movie is set. While coconut crabs are usually edible and even considered a delicacy in some places, the one in the movie was clearly not safe to eat. The liquid suggests the crab might have been rotting or sick, making it inedible. Coconut crabs are known for their strength and ability to crack open coconuts, but in this scene, it was more about survival gone wrong!

Q: What crustacean can "play dead" by remaining motionless for extended periods to avoid predators? A: The crayfish.

Q: Which crustacean builds complex burrows with multiple chambers for protection? A: The burrowing shrimp.

Q: What crustacean has the ability to "shed" parts of its body, such as claws, to escape predators? A: The crab.

Q: Which crustacean uses its claws to crush open mollusk shells for food? A: The green crab.

Q: What crustacean has specialized gills that allow it to extract oxygen from both water and air? A: The land crab.

Q: Which crustacean is known for its migratory behavior, traveling en masse to the ocean to spawn? A: The Christmas Island red crab.

Q: What crustacean is capable of producing a powerful strike with its claws, strong enough to break glass? A: The mantis shrimp.

Q: Which crustacean can live at extreme depths, sometimes over 7,000 feet underwater? A: The deep-sea amphipod.

Q: What crustacean is capable of creating "reefs" by attaching itself to hard surfaces and forming large colonies? A: The barnacle.

Q: Which crustacean has a highly specialized appendage called a "uropod," used for swimming? A: The shrimp.

Q: What crustacean is known for its symbiotic relationship with fish, often cleaning parasites from the fish's skin? A: The cleaner shrimp.

Q: Which crustacean produces a loud "popping" noise underwater by snapping its claw shut? A: The snapping shrimp.

Q: What crustacean is known for its ability to survive in low-oxygen environments by slowing its metabolism? A: The crayfish.

Q: Which crustacean can regrow its lost limbs over several molting cycles? A: The lobster.

Q: What crustacean can live in both marine and freshwater environments during different stages of its life? A: The amphipod.

Q: Which crustacean is known for "squirting" water to dig burrows in the sand? A: The razor clam (though technically not a crustacean, it's often associated with burrowing marine life).

Q: What crustacean uses its antennae to "smell" underwater and locate food? A: The shrimp.

Q: Which crustacean has the longest legs relative to body size of any arthropod? A: The Japanese spider crab.

Q: What crustacean has a unique appendage called a "pleopod," used for brooding eggs? A: The shrimp.

Q: Which crustacean is capable of switching its gender during its life? A: The slipper lobster.

Q: What crustacean has specialized "gill bailers" that help circulate water over its gills to aid in respiration? A: The crab.

Q: Which crustacean is known for its highly developed sense of taste and can detect chemicals in the water with its chemoreceptors? A: The crayfish.

Q: What crustacean species is known for carrying its food in specialized pouches under its abdomen? A: The amphipod.

Q: Which crustacean has a powerful tail fan used for rapid backward swimming to escape danger? A: The lobster.

Q: What crustacean can survive in the deep ocean with little sunlight, using bioluminescence to attract prey? A: The deep-sea copepod.

Q: Which crustacean uses its legs to "fan" oxygen-rich water over its eggs during the brooding process? A: The crayfish.

Q: What crustacean has a "double shell" exoskeleton, which molts as the animal grows? A: The barnacle.

Q: Which crustacean can survive on a diet primarily composed of wood and cellulose? A: The woodlouse (also called pill bug).

Q: What crustacean has a claw larger than its body, used to fight other males for mates? A: The fiddler crab.

Q: Which crustacean species is known for living in coral reefs and using coral for protection? A: The coral crab.

Q: What crustacean has a "telson" appendage that helps it navigate while swimming? A: The horseshoe crab (though technically not a crustacean, it shares many similarities).

Q: Which crustacean uses its claws to create "signals" in the water to communicate with others? A: The snapping shrimp.

Q: What crustacean is known for its ability to quickly burrow into mud or sand for protection? A: The mud crab.

Q: Which crustacean can live in hypersaline environments, such as salt lakes, where few other animals can survive? A: The brine shrimp.

Q: What crustacean is capable of walking on both the sea floor and on land? A: The coconut crab.

Q: Which crustacean species migrates in massive numbers, often creating "rivers" of crabs moving across the land? A: The Christmas Island red crab.

Q: What crustacean has been found to "herd" small schools of fish into traps using its pincers? A: The decorator crab.

Q: Which crustacean has a highly developed sense of hearing, capable of detecting low-frequency sound waves? A: The mantis shrimp.

Q: What crustacean can survive in the high-pressure environments of deep-sea trenches? A: The deep-sea amphipod.

Q: Which crustacean has a specialized carapace that allows it to blend into its environment, often covered in algae and sponges? A: The decorator crab.

Q: What crustacean is capable of producing a strong odor from its exoskeleton to deter predators? A: The green shore crab.

Q: Which crustacean can store oxygen in its blood, allowing it to survive for long periods without water? A: The land hermit crab.

Q: What crustacean can produce a foam-like substance to deter predators, making itself slippery and hard to grasp? A: The spiny lobster.

Q: Which crustacean is capable of moving in large swarms, sometimes covering large areas of the ocean floor? A: The Antarctic krill.

Q: What crustacean is known for its ability to survive in extreme environments, such as hydrothermal vents? A: The yeti crab.

Q: Which crustacean has specialized jointed limbs that allow it to "fold" its claws under its body for protection? A: The spiny lobster.

Q: What crustacean is known to "decorate" itself with pieces of seaweed and shells to camouflage from predators? A: The decorator crab.

Q: Which crustacean has been found to use tools, such as shells, to defend itself from predators? A: The hermit crab.

Q: What crustacean uses bioluminescent bacteria in its body to attract mates or scare off predators? A: The deep-sea shrimp.

Q: Which crustacean's exoskeleton is made of chitin, a substance that is also used in the production of biodegradable plastics? A: The lobster.

Q: What crustacean is capable of surviving in extremely acidic environments, such as volcanic lakes? A: The volcano shrimp.

Q: Which crustacean has the ability to flip itself upright by using its powerful tail, similar to a backflip? A: The lobster.

Q: What crustacean is known for its ability to molt its exoskeleton multiple times throughout its life? A: The blue crab.

Q: Which crustacean is known for migrating long distances to spawn, often returning to the same location each year? A: The Dungeness crab.

Q: What crustacean has specialized "grooming" appendages that it uses to clean its own body? A: The shrimp.

Q: Which crustacean can produce a toxic slime that deters predators and can cause skin irritation? A: The sea louse.

Q: What crustacean has been found to live symbiotically with anemones, using their stinging tentacles for protection? A: The porcelain crab.

Q: Which crustacean can "jump" by rapidly flexing its abdomen, launching itself off the sea floor? A: The sand shrimp.

Q: What crustacean is known to "fight" with other males by using its enlarged claws during mating season? A: The fiddler crab.

Q: Which crustacean has the ability to filter microscopic algae from the water as its primary food source? A: The krill.

Q: What crustacean is known for its symbiotic relationship with corals, often living inside coral colonies for protection? A: The coral crab.

Q: Which crustacean can survive by absorbing moisture from the air, allowing it to live far from water sources? A: The land crab.

Q: What crustacean can change the texture of its exoskeleton, becoming rough or smooth, depending on its surroundings? A: The ghost crab.

Q: Which crustacean species can live for extended periods in an anoxic (oxygen-free) environment? A: The brine shrimp.

Q: What crustacean is known for its bioluminescence, lighting up when disturbed by predators or during mating? A: The deep-sea ostracod.

Q: Which crustacean has a specialized mouthpart called a "mandible" used for crushing and grinding food? A: The lobster.

The horseshoe crab, despite its name, is not a true crab but rather a marine arthropod that has existed for over 450 million years. One of the most fascinating aspects of the horseshoe crab is its significant medical use. Its bright blue blood contains a unique substance called Limulus Amebocyte Lysate (LAL), which is used to test for bacterial contamination in medical devices, vaccines, and intravenous drugs.

The LAL test is highly sensitive to endotoxins, which are toxins released by certain bacteria. Even trace amounts of these endotoxins in medical products can cause severe reactions in humans, making LAL an essential tool in ensuring the safety of medical treatments. When horseshoe crab blood is exposed to bacterial endotoxins, it coagulates, providing a quick and effective way to detect contamination.

Horseshoe crabs are harvested for their blood, but efforts are made to collect it in a way that allows most crabs to survive and be released back into the wild. However, concerns about the sustainability of this practice have led to the development of synthetic alternatives, such as recombinant Factor C (rFC), which could eventually replace LAL testing while reducing the reliance on horseshoe crabs in the biomedical industry.

Advanced Fun Facts

Q: What is the largest species of terrestrial crustacean? A: The coconut crab (Birgus latro).

Q: What is the scientific term for the shedding of a crustacean's exoskeleton? A: Molting or ecdysis.

Q: Which crustacean is known for its ability to detect polarized light, providing enhanced underwater vision? A: The mantis shrimp (Stomatopoda).

Q: What structure do barnacles use to anchor themselves to surfaces? A: A cement-like gland located at the base of their shells.

Q: Which type of crustacean produces the bioluminescent compound luciferin? A: The ostracod (Cypridina).

Q: What specialized appendage do decapod crustaceans use for swimming? A: Pleopods.

Q: Which crustacean has the ability to regenerate lost appendages, including claws and legs? A: The blue crab (Callinectes sapidus).

Q: What is the name of the unique respiratory structure found in terrestrial crustaceans like woodlice? A: Pseudotrachea.

Q: Which crustacean is known for its "boxing" behavior, using sea anemones as gloves to defend itself? A: The boxer crab (Lybia tessellata).

Q: What deep-sea crustacean is known for farming bacteria on its setae for food? A: The yeti crab (Kiwa hirsuta).

Q: What term refers to the claws of crustaceans, such as those of crabs and lobsters? A: Chelipeds.

Q: Which crustacean has the most complex eyes, capable of detecting up to 12 different color channels? A: The mantis shrimp (Stomatopoda).

Q: What is the primary component of a crustacean's exoskeleton? A: Chitin.

Q: Which crustacean can migrate long distances across land to reach the ocean during breeding season? A: The Christmas Island red crab (Gecarcoidea natalis).

Q: What specialized structure do many marine crustaceans use to filter food from the water? A: Setal brushes on their maxillipeds.

Q: Which crustacean uses "bubbling" as a method to aerate its burrow and maintain humidity? A: The fiddler crab (Uca).

Q: What is the function of the uropods in many crustaceans? A: Uropods assist in swimming and steering.

Q: Which crustacean is capable of snapping its claws so fast that it creates a cavitation bubble that can stun prey? A: The snapping shrimp (Alpheidae).

Q: What is the name of the small, planktonic crustacean that is a major food source for whales? A: Krill (Euphausiacea).

Q: Which crustacean is known for its migratory swarms in response to seasonal changes in ocean currents? A: Antarctic krill (Euphausia superba).

Q: What is the name of the long, whip-like antennae used by crustaceans to sense their environment? A: Antennules.

Q: Which deep-sea crustacean species lives around hydrothermal vents and survives on bacteria in symbiosis? A: The yeti crab (Kiwa hirsuta).

Q: What species of crustacean is often parasitized by a barnacle that alters its reproductive behavior? A: The green crab (Carcinus maenas), parasitized by Sacculina.

Q: What pigment is responsible for the red color crustaceans develop when cooked? A: Astaxanthin.

Q: Which crustacean exhibits an unusual behavior where it builds "chimneys" around the entrance to its burrow? A: The ghost shrimp (Callianassa).

Q: What is the unique feature of the amphipod's body, distinguishing it from other crustaceans? A: Laterally compressed bodies.

Q: Which crustacean group has two distinct body segments: the cephalothorax and the abdomen? A: Decapods (e.g., crabs, lobsters, and shrimp).

Q: What is the primary method of locomotion for crabs? A: Walking sideways using their legs.

Q: What parasitic crustacean attaches to the tongues of fish, effectively replacing the organ? A: The tongue-eating louse (Cymothoa exigua).

Q: Which crustacean uses its tail to perform a rapid backward escape motion? A: The lobster (Homarus).

Q: What crustacean has an internal chamber called a "gas bladder," which helps control buoyancy? A: The pelagic red crab (Pleuroncodes planipes).

Q: What is the scientific term for the process by which crustaceans produce sound, often used for communication or defense? A: Stridulation.

Q: Which crustacean species produces a bioluminescent display during mating? A: The ostracod (Vargula hilgendorfii).

Q: What crustacean species is known for its role as a "cleaner," removing parasites from fish? A: The cleaner shrimp (Lysmata amboinensis).

Q: Which crustacean is known for its ability to survive and thrive in hypersaline environments? A: The brine shrimp (Artemia salina).

Q: What is the term for the hard plate that covers the cephalothorax of many crustaceans? A: Carapace.

Q: Which type of crustacean has a flattened body, allowing it to burrow into sandy or muddy substrates? A: The mole crab (Emerita).

Q: What is the scientific name for the class that includes all crustaceans? A: Malacostraca.

Q: Which crustacean forms colonies on rocks and ships, leading to biofouling? A: Barnacles (Cirripedia).

Q: What crustacean uses symbiotic relationships with anemones for protection, often carrying them on its back? A: The boxer crab (Lybia).

Q: Which crustacean can launch itself out of water, using its tail to escape predators? A: The sand flea (Talitridae).

Q: What is the term for the segmented appendages used by shrimp for walking and feeding? A: Pereiopods.

Q: Which crustacean is known for carrying its eggs externally in a specialized brood pouch? A: The amphipod.

Q: What crustacean is considered a "living fossil," having existed largely unchanged for hundreds of millions of years? A: The horseshoe crab (Limulus polyphemus).

Q: What term is used for the tiny planktonic crustaceans that make up a significant portion of marine biomass? A: Copepods.

Q: Which crustacean has the ability to crush shellfish with its oversized claw, exerting immense pressure? A: The stone crab (Menippe mercenaria).

Q: What parasitic crustacean attaches to the gills of fish and can damage fish populations? A: The sea louse (Lepeophtheirus salmonis).

Q: Which crustacean can "play dead" as a defense mechanism, lying motionless until predators pass? A: The crayfish.

Q: What crustacean can survive being frozen solid in ice and then revive when thawed? A: The fairy shrimp (Branchiopoda).

Q: Which deep-sea crustacean species is known for living in extreme environments, such as hydrothermal vents? A: The vent shrimp (Rimicaris exoculata).

Hungry? Here are some fun facts about edible crustaceans.

Edible crustaceans are a significant part of global cuisine, valued for their rich flavor and versatility. Some of the most commonly consumed crustaceans include crabs, lobsters, shrimp, and crayfish. These species are a major source of protein and are enjoyed in a variety of culinary traditions around the world.

Shrimp are among the most popular crustaceans, widely consumed in dishes like shrimp cocktails, stir-fries, and pastas. Their mild flavor and tender texture make them a favorite in many cuisines, especially in Asian, Mediterranean, and American dishes.

Lobsters are prized for their sweet, succulent meat and are often considered a luxury food. Lobster is typically boiled or steamed and served with butter, though it is also used in dishes like lobster rolls and bisques.

Crabs, including species like blue crabs and king crabs, are enjoyed for their delicate, slightly sweet meat. Crab meat is often used in salads, crab cakes, and soups.

Crayfish, also known as crawfish, are popular in Cajun and Creole cuisines, particularly in the southern United States. Crayfish boils, where the crustaceans are cooked with spices, corn, and potatoes, are a cultural staple.

Overall, edible crustaceans are a vital part of seafood diets, offering both nutritional benefits and delicious flavors.

In related news, my uncle once traveled to Louisiana and brought back an ice chest full of crayfish to our family reunion in Oklahoma. In Oklahoma, people use crayfish (we call them crawdads) for fish bait, not dinner.

Not wanting to look like a complete newbie, I welcomed the chance to eat crawdads.

As the steaming mountain of bright red crayfish hit the table, everyone dove in with seasoned expertise, cracking shells and popping juicy morsels into their mouths. Meanwhile, I grabbed my first crayfish, stared at it like it was a tiny alien, and awkwardly tried to figure out where to start. My uncle, noticing my struggle, chuckled and gave me a quick tutorial: "Twist the tail, suck the head."

Well, I did as I was told. I twisted that tail right off with all the finesse of a bulldozer, sending crayfish juice flying straight into my face. The whole table erupted in laughter as I wiped my eyes, and I still had no idea how to eat the thing! When it came time to "suck the head," I hesitated. "You want me to suck what now?"

Turns out, I ended up wearing more of the boil than I actually ate, but by the end of the night, I had mastered the art—or at least learned not to squirt everyone with crayfish juice. It was messy, but I left with a belly full of delicious food and a newfound respect for these little critters!

If you ever get the chance to share a pot of boiled crayfish with family and friends, dive in.

Mollusks

Mollusks are a large group of soft-bodied animals that live in many different environments, from the deep ocean to freshwater lakes and even on land. Some of the most common mollusks include snails, clams, octopuses, and squids. Although they look very different from each other, all mollusks have a few things in common.

Most mollusks have a soft body that's often protected by a hard shell. For example, clams and snails have shells that help keep them safe from predators. However, not all mollusks have shells. Octopuses and squids don't, but they are still part of the mollusk group because of other similarities in their body structure.

Mollusks also have a specialized foot they use for movement. In snails, the foot is the large, muscular part they slide along on, while in octopuses, their "foot" has evolved into arms and tentacles.

Another interesting feature of many mollusks is their mantle, which is a layer of tissue that can create the hard shell in species that have one. Mollusks also have a wide variety of feeding habits, from filter-feeding clams to predatory squids that hunt fish.

Overall, mollusks are incredibly diverse, living in many different habitats and playing important roles in their ecosystems.

Q: What is the largest mollusk species in the world? A: The giant squid (Architeuthis dux), which can grow up to 43 feet long.

Q: Which mollusk has a tongue-like structure covered with tiny teeth called a radula? A: The garden snail.

Q: What mollusk is known for producing pearls? A: The pearl oyster (Pinctada).

Q: Which type of mollusk can change its color to camouflage with its surroundings? A: The octopus.

Q: What mollusk has a shell that is divided into eight overlapping plates? A: The chiton.

Q: Which mollusk can squirt ink as a defense mechanism? A: The squid.

Q: What is the only mollusk that has been found in both freshwater and terrestrial environments? A: The freshwater snail.

Q: Which mollusk species has been known to "surf" on waves to move along the seafloor? A: The abalone.

Q: What is the largest known species of land snail? A: The giant African land snail (Achatina achatina).

Q: Which mollusk has the ability to regrow lost limbs, including arms or tentacles? A: The octopus.

Q: What species of mollusk uses jet propulsion by squirting water to escape predators? A: The cuttlefish.

Q: Which mollusk can close its shell tightly to trap water inside, allowing it to survive in dry conditions? A: The mussel.

Q: What mollusk has been known to drill into other mollusks' shells to feed on them? A: The moon snail.

Q: Which mollusk is capable of swimming by flapping its shell-like wings? A: The scallop.

Q: What mollusk is considered a delicacy and is often served as escargot? A: The common land snail (Helix pomatia).

Q: Which mollusk can produce up to 20,000 eggs at a time? A: The common octopus (Octopus vulgaris).

Q: What mollusk is known for its brightly colored and toxic mucus used to deter predators? A: The nudibranch.

Q: Which mollusk can use its siphon to burrow deep into sand or mud? A: The razor clam.

Q: What species of mollusk is known for its complex eyes, which can detect polarized light? A: The scallop.

Scallops are fascinating sea creatures known for their unique appearance and delicious taste. They have a fan-shaped shell, often brightly colored, with distinct ridges. Inside, the edible part is the tender, sweet muscle known as the "adductor," which helps the scallop open and close its shell. Scallops also have a series of small, bright blue or black eyes along the edge of their mantle, which help them detect movement and predators.

Scallops live in saltwater environments, usually on sandy or muddy sea floors. They're found in oceans around the world, from shallow coastal waters to deep-sea habitats.

Scallops are harvested either by diving or using dredges that drag along the ocean floor to collect them. As a delicacy, scallops are prized for their tender, slightly sweet flavor and are often served seared, grilled, or raw in dishes like ceviche.

Speaking of ceviche, what the heck is ceviche?

Ceviche is a popular seafood dish that originated in Latin America, particularly in coastal regions like Peru. It's made by marinating fresh, raw fish or seafood in citrus juice, usually lime or lemon. The acid in the citrus juice "cooks" the fish, giving it a firm texture and opaque appearance, even though no heat is used.

Ceviche is often mixed with ingredients like onions, cilantro, tomatoes, and chili peppers to give it a bright, fresh flavor. It's typically served cold and is especially popular in warm climates as a light, refreshing dish. It's a go-to in many Latin American countries, and there are different regional variations, but the concept remains the same — delicious, citrus-cured seafood!

Q: Which mollusk can change its body shape to fit into tight spaces and hide from predators? A: The octopus.

Q: What mollusk has a shell made up of two hinged parts? A: The clam.

Q: Which mollusk is known for its powerful foot that it uses to grip rocks in intertidal zones? A: The limpet.

Q: What mollusk has been observed using coconut shells as shelter? A: The veined octopus (Amphioctopus marginatus).

Q: Which mollusk produces a "love dart" during mating to increase reproductive success? A: The land snail.

Q: What mollusk is capable of producing its own light through bioluminescence? A: The firefly squid (Watasenia scintillans).

Q: Which mollusk can detect chemicals in the water to find food and mates using its tentacles? A: The sea slug.

Q: What mollusk is known for its ability to regenerate its shell after injury? A: The nautilus.

Q: Which mollusk has a protective internal shell called a "pen"?
A: The squid.

Q: What mollusk can "walk" on the ocean floor using its muscular foot? A: The sea cucumber.

Q: Which mollusk species is known to live inside discarded shells, such as those of other mollusks? A: The hermit crab (though technically a crustacean, it uses mollusk shells).

Q: What mollusk has a single, coiled shell and a large foot used for locomotion? A: The conch.

Q: Which mollusk uses a rasping radula to scrape algae off rocks for food? A: The periwinkle snail.

Q: What mollusk can squirt water from its siphon to defend itself from predators? A: The clam.

Q: Which mollusk can survive in oxygen-depleted environments by storing oxygen in its hemolymph? A: The marine mussel.

Q: What mollusk produces a highly toxic venom that can paralyze and kill prey? A: The cone snail.

Q: Which mollusk has a calcium carbonate shell that helps protect it from predators? A: The oyster.

Q: What mollusk can "jump" by forcefully closing its shell, creating propulsion? A: The scallop.

Q: Which mollusk uses its muscular foot to burrow into sand, mud, or sediment? A: The razor clam.

Q: What mollusk has the longest lifespan of any invertebrate, living up to 200 years? A: The ocean quahog (Arctica islandica).

Q: Which mollusk uses its tentacles to capture small fish and other prey? A: The cuttlefish.

Q: What mollusk is often used to study memory and learning because of its large neurons? A: The sea slug (Aplysia).

Q: Which mollusk has an open circulatory system, meaning blood flows freely within body cavities? A: The clam.

Q: What mollusk has a complex nervous system and is considered one of the most intelligent invertebrates? A: The octopus.

Q: Which mollusk is known for producing the substance nacre, which forms pearls? A: The pearl oyster.

Q: What species of mollusk can camouflage by changing the color and texture of its skin? A: The cuttlefish.

Q: Which mollusk has a brain-like ganglion structure and is capable of learning simple tasks? A: The octopus.

Q: What mollusk produces a sticky slime trail as it moves, which helps it glide over surfaces? A: The garden snail.

Q: Which mollusk secretes a byssus, a tough, thread-like structure used to attach itself to rocks? A: The mussel.

Q: What mollusk can inflate its body with water or air to appear larger to predators? A: The cuttlefish.

Q: Which mollusk has the ability to regenerate damaged parts of its shell? A: The abalone.

Q: What mollusk has a pair of complex eyes that can detect motion and light? A: The squid.

Q: Which mollusk can communicate with other members of its species using color changes? A: The octopus.

Q: What mollusk can drill through the shells of other mollusks to feed on them? A: The moon snail.

Q: Which mollusk can eject a cloud of ink to confuse predators and make a quick escape? A: The squid.

Q: What mollusk is capable of living in both freshwater and marine environments? A: The freshwater mussel.

Q: Which mollusk has been known to use tools, like coconut shells, for shelter? A: The veined octopus.

Q: What mollusk has the largest brain-to-body ratio of any invertebrate? A: The octopus.

Q: Which mollusk uses filter feeding to obtain its food, straining plankton from the water? A: The oyster.

Q: What mollusk has a tongue-like structure with rows of tiny teeth used to scrape food? A: The limpet.

Q: Which mollusk has a hard beak, similar to a parrot's, for eating prey? A: The octopus.

Q: What mollusk species is capable of producing a paralyzing toxin, dangerous even to humans? A: The cone snail.

Q: Which mollusk uses its powerful arms and suckers to grip prey? A: The giant Pacific octopus.

Q: What mollusk can survive for long periods without food by reducing its metabolic rate? A: The abalone.

Q: Which mollusk is capable of moving by expelling water through a siphon for jet propulsion? A: The squid.

Q: What mollusk uses a radula to drill into the shells of its prey? A: The whelk.

Q: Which mollusk produces a mucus to help it glide smoothly across surfaces? A: The snail.

Q: What mollusk can attach itself permanently to rocks using a byssus? A: The mussel.

Q: Which mollusk species has a coiled, external shell that helps protect it from predators? A: The conch.

Q: What mollusk uses a specialized "foot" to burrow and hide in sandy environments? A: The clam.

Q: Which mollusk has specialized cells called chromatophores that allow it to change colors rapidly? A: The octopus.

Q: What mollusk can live at great depths in the ocean and withstand high pressure? A: The giant squid.

Q: Which mollusk has eyes that can rotate independently, giving it a wide field of vision? A: The cuttlefish.

Q: What mollusk can close its shell tightly to protect itself from predators? A: The oyster.

Q: Which mollusk has the ability to use its tentacles to grip and manipulate objects? A: The octopus.

Q: What mollusk uses its radula to scrape algae off rocks for food? A: The limpet.

Q: Which mollusk uses its shell as a mobile home and will abandon it for a larger one as it grows? A: The hermit crab (uses mollusk shells).

Q: What mollusk produces a toxic mucus that can deter predators from eating it? A: The sea slug.

Q: Which mollusk uses its siphon to expel waste and control water flow in its body? A: The clam.

Q: What mollusk uses its muscular foot to move across the ocean floor in search of food? A: The abalone.

Q: Which mollusk species produces a special mucus that acts as a glue to stick to rocks and other surfaces? A: The limpet.

Q: What mollusk has a hinged shell that opens and closes to allow water in for feeding? A: The mussel.

Q: Which mollusk uses its tentacles to detect chemicals in the water and find prey? A: The cuttlefish.

Q: What mollusk produces a pearl as a defense mechanism against irritants that enter its shell? A: The oyster.

Oysters are a delicious treat, but knowing when they're safe to eat is crucial. A common rule of thumb is to only eat oysters in months that contain the letter "R" (September to April). This is because during warmer months (May to August), oysters are more likely to harbor harmful bacteria like Vibrio, which thrives in warm waters and can cause serious illness.

Historically, this old rule likely emerged because refrigeration wasn't available, so keeping oysters fresh in the heat was tricky. While modern refrigeration has improved food safety, it's still wise to be cautious.

And what does this have to do with the Ides of March? Well, just as Julius Caesar was warned to "beware the Ides of March" before his untimely assassination, perhaps we should be equally cautious when considering oysters outside the cooler months. No one wants their last meal to be a bad oyster!

Q: Which mollusk has a long, retractable siphon that it uses to breathe and feed? A: The clam.

Q: What mollusk has a lifespan of more than 100 years, making it one of the longest-lived animals? A: The ocean quahog.

Q: Which mollusk can change its skin texture to mimic the surrounding environment? A: The cuttlefish.

Q: What mollusk is capable of regrowing its arms or tentacles after losing them to predators? A: The octopus.

Some creatures have an amazing ability to regrow severed or dropped limbs, making them nature's ultimate survivors. One of the most well-known examples is the starfish (or sea star). If a starfish loses an arm, it can regenerate a whole new one—sometimes, they can even grow an entirely new body from a single arm!

Another famous regenerator is the axolotl, a type of salamander. Axolotls can regrow not only limbs but also parts of their heart, spinal cord, and even parts of their brain! They're like the superheroes of the animal world when it comes to regeneration.

Crabs and lobsters can also regrow their claws or legs if they lose them, though it takes a few molting cycles to fully replace the lost limb.

Geckos are pretty famous for their ability to drop their tails to escape predators and then regrow them later.

This ability is incredibly useful for survival, letting these animals bounce back from injuries that would be a lot more serious for most other creatures.

Q: Which mollusk can detach its foot as a defense mechanism to escape predators? A: The limpet.

Q: What mollusk uses its foot to cling to rocks in high-energy tidal zones? A: The barnacle.

Q: Which mollusk uses its sharp beak to crack open the shells of crabs and other prey? A: The octopus.

Octopuses have some pretty unique biology, and one of the coolest differences is their blood. Unlike humans and most other animals that have iron-based blood (hemoglobin), which makes it red, octopuses have copper-based blood (called hemocyanin), which gives their blood a bluish tint. This might seem strange, but there's a good reason for it. Copper is better at transporting oxygen in cold, low-oxygen environments, like the deep ocean where many octopuses live.

While iron-based hemoglobin works well for humans on land, where oxygen is plentiful, octopuses need something more efficient to thrive in their watery, sometimes low-oxygen world. Their blue blood helps them survive in extreme conditions, like deep ocean floors where oxygen can be scarce.

Q: What mollusk species has the ability to produce its own venom to immobilize prey? A: The cone snail.

Q: Which mollusk produces a sticky mucus trail that also serves as a scent trail for other mollusks? A: The snail.

Q: What mollusk is capable of releasing ink to create a smokescreen for escape? A: The squid.

Q: Which mollusk can detect vibrations in the water, helping it sense approaching predators? A: The clam.

Q: What mollusk uses its tentacles to catch small fish, crabs, and other marine life? A: The octopus.

Q: Which mollusk can survive in both brackish and freshwater environments? A: The freshwater clam.

Q: What mollusk produces a series of clicks or popping sounds during mating rituals? A: The firefly squid.

Squid and octopuses may seem similar at first glance, but they've got some key differences. Both are part of the cephalopod family, but their body shapes and behaviors set them apart.

Octopuses have round, squishy bodies and eight arms, with each arm lined with suction cups. They're usually loners, hanging out in dens, and are known for being super smart— solving puzzles and escaping tight spots like pros! They don't have any bones, so they can squeeze into all sorts of places.

Squid, on the other hand, are more streamlined. They have a torpedo-shaped body, with eight arms plus two longer tentacles used for grabbing prey. Unlike octopuses, squid tend to live in groups and are faster swimmers, jetting around the ocean. They also have a tiny internal shell called a "pen" for structure.

In short, octopuses are like the brainy, flexible introverts, while squid are the speedy, social go-getters of the ocean!

Q: Which mollusk uses a special muscular foot to move along surfaces in search of food? A: The snail.

Q: What mollusk has a hard outer shell that protects its soft body from predators? A: The abalone.

Q: Which mollusk produces a secretion that helps it adhere to rocks and other hard surfaces in the ocean? A: The barnacle.

Advanced Fun Facts

Q: What protein is responsible for the formation of a mollusk's shell? A: Conchiolin.

Q: Which group of mollusks is known to have developed torsion, a unique twisting of the body during development? A: Gastropods.

Q: What is the name of the specialized feeding organ found in many mollusks that is used to scrape or cut food? A: Radula.

Q: Which class of mollusks lacks a radula and instead uses filter feeding to obtain nutrients? A: Bivalvia (e.g., clams, mussels, oysters).

A sponge is a simple, aquatic animal that belongs to the phylum Porifera. They are some of the most basic multicellular organisms on Earth, and they have no organs, nerves, or muscles. Instead, sponges have a porous body structure, which allows them to filter water through their bodies to capture food particles, such as bacteria and plankton. Sponges come in many shapes, sizes, and colors, and they play an important role in marine ecosystems by filtering large amounts of water.

Sponges are primarily found in marine environments, from shallow coastal waters to the deep sea, though some species live in freshwater. Despite their simple structure, sponges have been around for over 500 million years, making them some of the oldest animals on Earth!

Q: What is the primary respiratory structure used by most aquatic mollusks to extract oxygen from water? A: Gills, or ctenidia.

Q: Which group of mollusks has developed a highly complex nervous system and demonstrates problem-solving abilities? A: Cephalopods (e.g., octopuses, squids).

Q: What is the term for the internal, chambered shell structure found in species like the nautilus? A: Siphuncle.

Q: Which mollusk group is known for having an open circulatory system, where hemolymph bathes the organs directly? A: Most gastropods and bivalves.

Q: What is the main role of the mollusk's mantle? A: To secrete the shell and protect internal organs.

Q: Which cephalopod is known for producing a specialized pigment called sepia, used in ink for defense? A: The common cuttlefish (Sepia officinalis).

Q: How do octopuses achieve such remarkable camouflage, adjusting both color and texture? A: Through chromatophores and papillae, which allow changes in skin color and texture.

Q: What is the term for the free-swimming larval stage of many mollusks, particularly gastropods and bivalves? A: Veliger.

Q: Which mollusk has developed the strongest bite of any animal relative to its body size, capable of breaking shells? A: The octopus.

Q: What structure in cephalopods is homologous to the foot of other mollusks and has evolved into arms and tentacles? A: The siphon or funnel.

Q: Which ancient mollusk group is known for its external shell and has existed for hundreds of millions of years? A: Nautiloids.

Q: What is the name of the specialized cells in cephalopods that can change color by expanding or contracting pigment sacs? A: Chromatophores.

Q: How do bivalves, like oysters, produce pearls? A: By secreting layers of nacre around an irritant inside their shell.

Q: What is the significance of the osphradium in aquatic gastropods? A: It is a sensory organ used to detect water quality and chemical changes in the environment.

Q: What is the unique respiratory adaptation of pulmonate gastropods that allows them to live on land? A: They have evolved a lung-like structure for breathing air, instead of gills.

Q: In cephalopods, what is the primary function of the siphon? A: It is used for jet propulsion, expelling water to move the animal rapidly.

Q: Which gastropod is known to have harpoon-like radular teeth that can inject venom into prey? A: The cone snail (Conus species).

Q: What is the purpose of the operculum in many gastropods? A: It acts as a "trapdoor," sealing the shell's opening for protection.

Q: How do chitons use their radula differently from other mollusks? A: Their radula is fortified with iron to scrape algae off rocks.

Q: Which bivalve can swim by clapping its shells together rapidly? A: The scallop.

Q: What is the function of the byssal threads in bivalves like mussels? A: They are used to anchor the mussel to solid surfaces.

Q: How do cephalopods achieve rapid changes in texture, such as spiked or smooth surfaces, for camouflage? A: They use muscular hydrostats called papillae.

Q: What is the name of the internal, spiral-shaped shell in squid that is a vestigial remnant of an ancestral external shell? A: The pen or gladius.

Q: Which group of mollusks is capable of bioluminescence, producing light to communicate or distract predators? A: Certain cephalopods, such as the firefly squid.

Q: What is the term for the mollusk's unique embryonic twisting process that results in the rotation of internal organs in gastropods? A: Torsion.

Q: Which deep-sea mollusk species has a symbiotic relationship with bacteria that help it metabolize sulfur compounds from hydrothermal vents? A: The scaly-foot gastropod (Chrysomallon squamiferum).

Q: What adaptation allows land snails to retain moisture and prevent desiccation in terrestrial environments? A: They secrete mucus and can retract into their shells, closing the opening with an operculum.

Q: How does the radula of carnivorous mollusks, like the moon snail, differ from that of herbivorous species? A: It is often specialized with sharp, cutting teeth to drill into prey.

Q: Which mollusk is known for being a keystone species in ecosystems, heavily influencing the structure of the community by its grazing habits? A: The limpet.

Q: What is the specialized larval form unique to cephalopods that is distinct from other mollusk larvae? A: They do not have a distinct larval form; young hatch as miniature adults.

Q: What part of the squid's anatomy is known to contain some of the largest axons (nerve fibers) found in the animal kingdom? A: The giant axon in the squid's nervous system, which controls its mantle and jet propulsion.

Q: What is the role of statocysts in mollusks like cephalopods and gastropods? A: They help the animal maintain balance and orientation.

Q: Which mollusk group is the only one to have both an external shell and an internal siphuncle that allows them to control buoyancy? A: Nautiloids.

Q: What environmental change led to the evolution of air-breathing in pulmonate gastropods? A: The colonization of terrestrial environments, where gills became inefficient for gas exchange.

Q: What is the main distinguishing feature between prosobranch and opisthobranch gastropods? A: The position of the gills and mantle cavity — prosobranchs have them in front of the heart, while opisthobranchs have them behind.

Q: How do giant clams obtain most of their nutrition? A: They host symbiotic algae (zooxanthellae) in their tissues, which provide nutrients through photosynthesis.

Q: What mollusk has a beak similar to a bird's, used to tear apart prey? A: The octopus.

Q: Which gastropod species has evolved to be completely shell-less and exhibits bright warning colors as a defense mechanism? A: Nudibranchs.

Q: What unique feature does the argonaut, a species of octopus, have that distinguishes it from other cephalopods? A: Females produce a thin, paper-like shell to protect their eggs.

Q: How do cephalopods achieve their highly coordinated and rapid movements? A: Through a complex nervous system and coordination between their mantle muscles and siphon.

Q: What is the ecological significance of bivalves like oysters in marine ecosystems? A: They are filter feeders that improve water quality and provide habitat structure for other marine organisms.

Q: Which gastropod is known for having a venomous harpoon-like tooth that can deliver a lethal sting to fish and even humans? A: The cone snail (Conus).

Q: What is the purpose of the "beak" found in cephalopods like squids and octopuses? A: It is used to tear apart prey into smaller, digestible pieces.

Q: What is the term for the hardened plate at the entrance of a bivalve shell that acts as a defense mechanism? A: The hinge ligament or teeth.

Q: What function does the peristomal groove serve in mollusks like chitons? A: It channels food particles toward the mouth for feeding.

Q: How do gastropods like abalones produce their highly iridescent inner shell layer, known as mother-of-pearl? A: Through the secretion of nacre, a combination of calcium carbonate and conchiolin.

While living in Alaska as a child, we often looked forward to a trip to Seward, Alaska for clamming season.

Clamming in Seward, Alaska, is an awesome adventure for anyone who loves the outdoors. It's all about heading to the beach during low tide, digging in the sand, and pulling up clams. In Seward, razor clams are the prize, and they're not too hard to find if you know what you're doing. The beaches, especially along the Kenai Peninsula, are perfect for clamming, and you'll often see people armed with shovels and buckets, digging away.

But be advised. Don't show up with your ordinary shovel. A clamming shovel has a long narrow face, and is typically short handled. Also, check the tide schedule. The tides around Seward are some of the fastest in the world.

Wee Beasties

Thanks to Claire Fraser from the amazing series *Outlander* for this "wee beasties" insight on the world of bacteria and viruses, and while Claire was a time traveller from the World War II era to a time before the 19th century, it was not until the 1800s that Louis Pasteur opened our eyes to the world of wee beasties.

Louis Pasteur was a pioneering scientist who made groundbreaking contributions to the field of microbiology, fundamentally changing our understanding of germs, bacteria, and disease. In the mid-19th century, Pasteur conducted experiments that debunked the prevailing theory of spontaneous generation—the belief that life could arise from non-living matter. Through a series of meticulous experiments using swan-necked flasks, Pasteur showed that microorganisms, such as bacteria, were present in the air and did not spontaneously appear.

Louis Pasteur made several key discoveries throughout his career, spanning from the 1850s to the 1880s. Here are the main dates associated with his groundbreaking work:

- 1857: Pasteur began studying fermentation and discovered that microorganisms were responsible for the process. This led to his realization that microbes play a role in chemical changes.
- 1861: Pasteur conducted his famous swan-neck flask experiment, disproving the theory of spontaneous generation. This experiment showed that microorganisms come from the air, not from non-living matter.
- 1864: He developed the process of pasteurization, where liquids like wine and milk are heated to kill harmful bacteria. This discovery had a huge impact on food safety.
- 1877–1881: Pasteur turned his attention to infectious diseases and identified specific bacteria as the cause of diseases like anthrax. In 1881, he successfully created the anthrax vaccine, proving that weakened pathogens could be used to immunize animals.
- 1885: Pasteur developed the first rabies vaccine and successfully administered it to a young boy who had been bitten by a rabid dog, saving his life. This marked one of his most famous achievements.

Wee beasties like germs, bacteria, and viruses don't fit neatly into the traditional Linnaean taxonomy, which was originally designed to classify complex life forms like plants and animals. However, here's how they relate to the system:

- Bacteria: These are classified within the Domain Bacteria, which is one of the three domains in modern taxonomy (the other two being Archaea and Eukarya). Within this domain, bacteria are further classified into phyla, classes, orders, families, genera, and species—just like plants and animals.

For example, the common gut bacterium Escherichia coli belongs to the genus Escherichia and species coli.
- Viruses: Viruses don't fit into Linnaean taxonomy at all because they aren't considered living organisms. Instead, they fall under virology and are classified based on factors like their genetic material (DNA or RNA), shape, and how they replicate. Examples include the influenza virus and SARS-CoV-2.
- Germs: The term "germs" is a general, non-scientific term that refers to any microorganisms that can cause disease. This includes bacteria, viruses, fungi, and protozoa, so "germs" span across different categories in taxonomy.

In short, bacteria have a place in the Linnaean system, viruses do not, and "germs" is just a broad term for microorganisms in general. Given the theme of this book, the following trivia questions and answers focuses on the amazing world of bacteria.

Q: What percentage of the human body's cells are bacterial rather than human? A: About 90% of the cells in the human body are bacterial.

Q: Which bacteria are commonly used in yogurt production? A: Lactobacillus and Streptococcus species.

Q: What shape are Bacillus bacteria? A: Rod-shaped.

Q: What is the name of the bacteria responsible for strep throat? A: Streptococcus pyogenes.

Q: Which bacteria can survive extreme heat and are found in hot springs? A: Thermophiles.

Q: What bacterial species causes Lyme disease? A: Borrelia burgdorferi.

Q: Which bacteria glow in the dark and are bioluminescent? A: Vibrio fischeri.

Q: What type of bacteria causes tuberculosis? A: Mycobacterium tuberculosis.

Q: Which bacteria can convert nitrogen in the air into a form that plants can use? A: Rhizobium.

Q: What bacterial infection is often associated with food poisoning from undercooked chicken? A: Salmonella.

Fortunately, not all bactria leads to deadly forms of food poisoning.

Bacteria might sound like something you want to avoid, but they actually play a huge role in making some of our favorite foods! Many bacteria are essential for food production, and without them, we wouldn't have classics like yogurt, cheese, or even pickles.

Take yogurt for example—bacteria like Lactobacillus and Streptococcus are added to milk, where they ferment the sugars and give yogurt its tangy flavor and thick texture. Similarly, cheese production relies on different bacteria to break down the milk proteins and fats, creating a wide variety of flavors and textures, from soft brie to sharp cheddar.

Then there's sourdough bread, which gets its unique taste from naturally occurring lactic acid bacteria. These bacteria work alongside yeast during fermentation, giving the bread its distinctive tang.

Fermented foods like sauerkraut, kimchi, and pickles are also thanks to bacteria. In these cases, bacteria help preserve the vegetables and create that delicious sour flavor by producing lactic acid.

Beyond flavor, bacteria are also good for our health, helping to introduce probiotics—good bacteria that benefit our gut. So, next time you enjoy a slice of cheese or a cup of yogurt, remember it's bacteria working behind the scenes to make it delicious!

Q: What bacteria is responsible for causing tetanus? A: Clostridium tetani.

Q: Which bacteria play a key role in breaking down oil spills in the ocean? A: Alcanivorax borkumensis.

Q: What is the main bacterium responsible for causing cavities in teeth? A: Streptococcus mutans.

Q: What bacteria are commonly found in the intestines and aid in digestion? A: Escherichia coli (E. coli).

Q: Which bacterial infection is characterized by a severe cough and is known as whooping cough? A: Bordetella pertussis.

Q: What bacteria cause cholera, leading to severe dehydration from diarrhea? A: Vibrio cholerae.

Q: Which bacteria are used in sewage treatment to break down organic matter? A: Aerobic bacteria such as Pseudomonas.

Bacteria can be tiny troublemakers when it comes to spreading disease. These microscopic organisms are everywhere—on surfaces, in the air, in water, and even on our skin. While many bacteria are harmless or even helpful, some can make us sick by spreading infections in various ways.

One of the most common ways bacteria spread is through person-to-person contact. For example, if someone with strep throat sneezes or coughs, the bacteria can become airborne, and if you breathe it in, you might get sick too. Bacteria also spread through touching contaminated surfaces—things like door handles, phones, or countertops. If you touch one of these surfaces and then touch your face, you can transfer the bacteria to yourself.

Bacteria also spread through food and water. When food isn't properly cooked or handled, bacteria like Salmonella or E. coli can infect you. Drinking contaminated water can introduce harmful bacteria into your system too.

Insects, like mosquitoes or ticks, can also spread bacteria. For example, Lyme disease is caused by bacteria carried by ticks.

The good news? Washing your hands, cooking food properly, and staying clean can prevent most bacterial infections from spreading. Simple hygiene goes a long way in keeping those pesky germs at bay!

Q: What bacteria can cause a dangerous skin infection known as necrotizing fasciitis? A: Streptococcus pyogenes.

Q: Which bacteria are known for their corkscrew shape and ability to move through viscous environments? A: Spirochetes.

Q: What bacteria are responsible for the fermentation process in sourdough bread? A: Lactobacillus species.

Q: Which bacteria are commonly found on the skin and can cause acne? A: Propionibacterium acnes.

Q: What bacteria can survive radiation doses thousands of times higher than what would kill a human? A: Deinococcus radiodurans.

Q: Which bacteria are involved in making vinegar through the fermentation of alcohol? A: Acetobacter.

Q: What bacteria live in symbiosis with cows, helping them digest cellulose in grass? A: Ruminococcus species.

Q: Which bacteria can photosynthesize, contributing oxygen to Earth's atmosphere? A: Cyanobacteria.

Q: What bacterial species is responsible for causing botulism? A: Clostridium botulinum.

Q: Which bacteria can convert lactose into lactic acid, helping in the production of dairy products? A: Lactobacillus bulgaricus.

Q: What bacteria produce the antibiotic streptomycin? A: Streptomyces species.

Q: Which bacteria are known to form biofilms on surfaces like teeth or medical implants? A: Pseudomonas aeruginosa.

Q: What bacteria cause syphilis, a sexually transmitted infection? A: Treponema pallidum.

Q: Which bacteria are known for their ability to fix nitrogen in the roots of legumes? A: Rhizobium species.

Q: What bacteria can live inside icebergs and other cold environments, thriving at freezing temperatures? A: Psychrophiles.

Q: Which bacteria were the first to be genetically modified in the lab to produce human insulin? A: Escherichia coli (E. coli).

Q: What bacteria are commonly associated with stomach ulcers? A: Helicobacter pylori.

Q: Which bacteria are used in bioremediation to clean up heavy metals from contaminated soils? A: Geobacter species.

Q: What bacteria have a protective outer structure called an endospore, allowing them to survive extreme conditions? A: Bacillus and Clostridium species.

Q: Which bacteria help produce methane gas in the stomachs of cows and other ruminants? A: Methanobrevibacter.

Q: What bacteria are responsible for causing leprosy? A: Mycobacterium leprae.

Ruminant animals, like cows, sheep, and goats, produce a lot of methane because of the unique way they digest their food. These animals have a specialized stomach called a rumen, where food is broken down by microbes, including bacteria, rather than just by stomach acid.

Ruminants mainly eat tough, fibrous plants like grass, which are difficult to digest. To break down these plant fibers, ruminants rely on the bacteria in their rumen to ferment the food. During this fermentation process, certain bacteria produce hydrogen and carbon dioxide as byproducts. Other bacteria, called methanogens, use these gases to produce methane, which is then expelled by the animal, mostly through burping.

Methane production isn't directly beneficial to the animal itself, but it is a natural part of the fermentation process that allows ruminants to get nutrients from plants that most other animals can't digest. The bacteria help break down cellulose, which provides the animal with energy and nutrients that would otherwise be locked away in the tough plant material.

So while the methane doesn't benefit the animal directly, the bacterial fermentation process is essential for ruminants to thrive on plant-based diets. The methane is just a byproduct of that process.

Q: Which bacteria are used to make sauerkraut through lactic acid fermentation? A: Lactobacillus plantarum.

Q: What bacteria cause diphtheria, a serious throat infection? A: Corynebacterium diphtheriae.

Q: Which bacteria can break down cellulose and are used in composting to decompose plant material? A: Cellulomonas species.

Q: What bacterial species produce antibiotics such as tetracycline? A: Streptomyces.

Q: Which bacteria can be found in the mouth and are responsible for bad breath (halitosis)? A: Fusobacterium.

Q: What bacterial species are commonly used in probiotics to promote gut health? A: Bifidobacterium and Lactobacillus species.

Q: Which bacteria are involved in the formation of coral reefs, providing them with essential nutrients? A: Cyanobacteria (symbiotic with corals).

Q: What bacteria are responsible for breaking down dead organic material in soil and water? A: Decomposing bacteria like Actinobacteria.

Q: Which bacterial species are used to produce biofuels from plant waste? A: Clostridium thermocellum.

Q: What bacteria can form iron oxide deposits, creating "rust" underwater in shipwrecks? A: Gallionella species.

Q: Which bacteria are known to form symbiotic relationships with termites, helping them digest wood? A: Spirochaetes.

Q: What bacteria are involved in cheese production, giving certain cheeses their distinct flavors? A: Lactococcus lactis.

Q: Which bacteria are known for their spiral shape and are associated with causing foodborne illnesses from undercooked poultry? A: Campylobacter jejuni.

Q: What bacteria can produce deadly toxins, causing anthrax? A: Bacillus anthracis.

Q: Which bacteria are important for the nitrogen cycle by converting ammonia into nitrites? A: Nitrosomonas.

Q: What bacteria are capable of photosynthesis without producing oxygen, using sulfur instead of water? A: Purple sulfur bacteria.

Q: Which bacteria are found in yogurt and contribute to its sour taste and probiotic benefits? A: Lactobacillus delbrueckii.

Q: What bacterial species can survive extreme acid conditions, thriving in volcanic environments? A: Acidophiles.

Q: Which bacteria are used in the industrial production of vinegar? A: Acetobacter aceti.

Q: What bacteria can cause food poisoning through the production of toxins in improperly canned foods? A: Clostridium botulinum.

Q: Which bacteria are responsible for causing the Black Death (bubonic plague)? A: Yersinia pestis.

Q: What bacteria are found in soil and play an essential role in the nitrogen cycle by converting nitrogen into a usable form for plants? A: Azotobacter.

Q: Which bacteria can ferment lactose, producing lactic acid as a byproduct in dairy products? A: Streptococcus thermophilus.

Q: What bacteria are known to form symbiotic relationships with bioluminescent fish and help them glow? A: Vibrio species.

Q: Which bacteria are responsible for causing Legionnaires' disease, a severe form of pneumonia? A: Legionella pneumophila.

Q: What bacteria can survive and grow in extreme salt concentrations, such as in salt lakes? A: Halophiles.

Q: Which bacteria are capable of reducing sulfur to produce hydrogen sulfide, giving swamp mud its characteristic smell? A: Sulfate-reducing bacteria like Desulfovibrio.

Q: What bacterial species is known to help digest dead plants and other organic material in wetlands? A: Clostridium species.

Q: Which bacteria are responsible for converting nitrites into nitrates in the soil, playing a key role in the nitrogen cycle? A: Nitrobacter.

Q: What bacteria can withstand extreme heat and pressure, living near deep-sea hydrothermal vents? A: Thermophiles like Pyrolobus fumarii.

Q: Which bacteria are responsible for causing typhoid fever? A: Salmonella typhi.

Q: What bacteria are involved in the process of methanogenesis, producing methane as a byproduct? A: Methanogenic archaea, such as Methanobrevibacter.

Here are a few hairsplitting and geeky fun facts regrding fungi and bacteria. Feel free to skip over this section if you are allergic to big words.

Fungi and bacteria are both microorganisms, but they are quite different in terms of structure, function, and behavior. Here's a breakdown of their main differences:

- Cell Type: Bacteria are prokaryotic, meaning they don't have a nucleus or other membrane-bound organelles. Their genetic material floats freely in the cell. Fungi are eukaryotic, meaning they have a nucleus and organelles (like mitochondria), making their cells more complex.
- Size: Bacteria are typically much smaller than fungi. Bacteria are usually about 1-10 micrometers, while fungi, especially multicellular ones like molds and mushrooms, can be much larger.

- Structure: Bacteria are usually single-celled organisms with simple shapes (spherical, rod-shaped, or spiral). Fungi can be single-celled (like yeast) or multicellular (like molds and mushrooms) and have a more complex structure, often forming long, thread-like strands called hyphae.
- Reproduction: Bacteria primarily reproduce by binary fission, where one cell splits into two. Fungi reproduce using spores, which can be sexual or asexual, depending on the species.
- Nutrition: Bacteria can be autotrophic (make their own food, like photosynthetic bacteria) or heterotrophic (consume organic material). Fungi are exclusively heterotrophic, meaning they absorb nutrients from organic material, often breaking down dead organisms (decomposers).
- Cell Wall: Bacterial cell walls are made of peptidoglycan. Fungal cell walls are made of chitin.

In summary, fungi are more complex, larger organisms with a eukaryotic cell structure, while bacteria are simpler, smaller prokaryotes. Both play essential roles in ecosystems, but they differ greatly in how they are built and function.

Q: Which bacteria can survive and thrive in radioactive environments? A: Deinococcus radiodurans.

Q: What bacteria are commonly used in genetic engineering to insert foreign DNA into plants? A: Agrobacterium tumefaciens.

Q: Which bacteria cause peptic ulcers in humans? A: Helicobacter pylori.

Q: What bacterial species produces the antibiotic penicillin? A: Penicillium (a fungus, but the bacteria that penicillin targets are Staphylococcus and Streptococcus).

Q: Which bacteria are commonly used in the fermentation of pickles and sauerkraut? A: Leuconostoc species.

Q: What bacteria are used in wastewater treatment to break down organic waste? A: Aerobic bacteria like Pseudomonas.

Q: Which bacteria are known for their ability to cause food spoilage by breaking down proteins in meat? A: Pseudomonas species.

Q: What bacterial species is commonly found in raw or undercooked pork and can cause trichinosis? A: Trichinella spiralis (though not a bacteria, it is caused by a parasitic nematode, commonly associated with bacterial contamination).

Q: Which bacteria are found in dental plaque and contribute to tooth decay? A: Streptococcus mutans.

Q: What bacteria are known to convert organic waste into biogas in anaerobic digesters? A: Methanogens like Methanosaeta.

Q: Which bacteria are responsible for the disease anthrax? A: Bacillus anthracis.

Q: What bacteria can thrive in the acidic environment of a human stomach? A: Helicobacter pylori.

Q: Which bacteria are known for their resistance to antibiotics and are a major cause of hospital-acquired infections? A: Methicillin-resistant Staphylococcus aureus (MRSA).

Q: What bacteria cause the deadly disease cholera? A: Vibrio cholerae.

Q: Which bacteria can grow on oil spills and help break down hydrocarbons? A: Alcanivorax borkumensis.

Q: What bacteria cause Legionnaires' disease? A: Legionella pneumophila.

Q: Which bacteria thrive in low-oxygen environments and produce methane as a byproduct? A: Methanobacteria.

Q: What bacteria are used in the production of antibiotics such as erythromycin? A: Saccharopolyspora erythraea.

Q: Which bacteria are known to cause severe skin infections like cellulitis? A: Staphylococcus aureus.

Q: What bacteria cause leprosy? A: Mycobacterium leprae.

Q: Which bacteria are responsible for causing Lyme disease? A: Borrelia burgdorferi.

Q: What bacteria produce acetic acid as a byproduct of fermentation? A: Acetobacter aceti.

Q: Which bacteria are used to ferment cocoa beans into chocolate? A: Lactobacillus and Acetobacter species.

Q: What bacteria are involved in the decomposition of dead organic matter in compost? A: Bacillus and Clostridium species.

Q: Which bacteria are found in the gut and are critical for producing vitamins like B12 and K? A: Bacteroides and Escherichia coli (E. coli).

Q: What bacteria are involved in fixing nitrogen in the roots of legumes? A: Rhizobium.

Q: Which bacteria produce biofilms, which can coat medical devices and cause infections? A: Pseudomonas aeruginosa.

Q: What bacteria are responsible for the fermentation of soy sauce? A: Lactobacillus and Aspergillus oryzae.

Q: Which bacteria are commonly used in the production of kombucha? A: Acetobacter and Gluconobacter species.

Q: What bacteria can withstand and even thrive in radioactive environments? A: Deinococcus radiodurans.

Finally, let's part on something positive.

Microbiological organisms, like bacteria and fungi, aren't just tiny troublemakers—they actually play a huge role in keeping us healthy. The most obvious example is the gut microbiome. Our digestive system is home to trillions of beneficial bacteria that help break down food, produce vitamins (like B12 and K), and keep harmful bacteria in check. Without these good guys, our digestion would be a mess, and our immune system would struggle.

Probiotics, which are live beneficial bacteria found in foods like yogurt or supplements, are another great example. They help balance out the gut and improve digestion, boost immunity, and even reduce the risk of certain infections. Having a healthy gut flora is crucial for overall well-being.

Another big win from microbiology is in the field of antibiotics. Many antibiotics, like penicillin, are derived from bacteria or fungi. These drugs revolutionized medicine by giving us a way to fight off bacterial infections that used to be deadly.

And don't forget vaccines. Some vaccines use weakened or dead microorganisms to teach our immune system how to recognize and fight diseases.

In short, microbiological organisms are essential in medicine — from keeping our gut in check to saving lives with antibiotics and vaccines. They're the unsung heroes of health!

And thanks also to Claire Fraser. We love you and your wee beasties.

www.ingramcontent.com/pod-product-compliance
Lightning Source LLC
Chambersburg PA
CBHW052147220526
45471CB00004B/1563